Communications
in Computer and Information Science 1162

Commenced Publication in 2007
Founding and Former Series Editors:
Phoebe Chen, Alfredo Cuzzocrea, Xiaoyong Du, Orhun Kara, Ting Liu,
Krishna M. Sivalingam, Dominik Ślęzak, Takashi Washio, Xiaokang Yang,
and Junsong Yuan

More information about this series at http://www.springer.com/series/7899

Greg H. Parlier · Federico Liberatore ·
Marc Demange (Eds.)

Operations Research and Enterprise Systems

8th International Conference, ICORES 2019
Prague, Czech Republic, February 19–21, 2019
Revised Selected Papers

 Springer

Editors
Greg H. Parlier
INFORMS
Catonsville, MD, USA

Federico Liberatore
Universidad Carlos III de Madrid
Madrid, Spain

Marc Demange
RMIT University
Melbourne, VIC, Australia

ISSN 1865-0929   ISSN 1865-0937   (electronic)
Communications in Computer and Information Science
ISBN 978-3-030-37583-6   ISBN 978-3-030-37584-3   (eBook)
https://doi.org/10.1007/978-3-030-37584-3

This Springer imprint is published by the registered company Springer Nature Switzerland AG
The registered company address is: Gewerbestrasse 11, 6330 Cham, Switzerland

Preface

This book includes extended and revised versions of selected papers from the 8th International Conference on Operations Research and Enterprise Systems (ICORES 2019), held in Prague, Czech Republic during February 19–21, 2019.

ICORES 2019 received 80 paper submissions from 35 countries, of which 11% are included in this book. These papers were selected based on several criteria including reviews provided by Program Committee members, session chair assessments, and also program chair perspectives across all papers included in the technical program. The authors of these selected papers were then invited to submit revised and extended versions of their papers for formal publication.

The purpose of annual ICORES is to bring together researchers, engineers, and practitioners interested in both advances and applications in the field of Operations Research. Two simultaneous tracks were held, one covering domain independent methodologies and technologies, and the other practical work developed in specific application areas.

The papers selected for this book contribute to Operations Research and our better understanding of complex Enterprise Systems. We commend each of the authors for their contributions, and gratefully thank our many reviewers who ensured high quality for this publication.

February 2019

Greg H. Parlier
Federico Liberatore
Marc Demange

Organization

Conference Chair

Marc Demange   RMIT University, Australia

Program Co-chairs

Greg H. Parlier   NCSU, USA
Federico Liberatore   Cardiff University, UK

Program Committee

Bernardetta Addis	Université de Lorraine, France
El-Houssaine Aghezzaf	Ghent University, Belgium
Javier Alcaraz	Universidad Miguel Hernandez de Elche, Spain
Lionel Amodeo	University of Technology of Troyes, France
Necati Aras	Bogazici University, Turkey
Eduardo Barbosa	Brazilian National Institute for Space Research (INPE), Brazil
Patrizia Beraldi	University of Calabria, Italy
Lotte Berghman	Toulouse Business School, France
David Bergman	University of Connecticut, USA
Gianpiero Bianchi	Italian National Institute of Statistics Istat, Italy
Giancarlo Bigi	University of Pisa, Italy
Cyril Briand	LAAS-CNRS, France
Renato Bruni	University of Roma La Sapienza, Italy
Ahmed Bufardi	École Polytechnique Fédérale de Lausanne, Switzerland
Sujin Bureerat	KhonKaen University, Thailand
Valentina Cacchiani	University of Bologna, Italy
Mirko Cesarini	University of Milan Bicocca, Italy
Bo Chen	University of Warwick, UK
Il-Gyo Chong	Samsung Electronics, South Korea
Andy Chow	City University of Hong Kong, Hong Kong, China
Andre Cire	University of Toronto, Canada
Roberto Cordone	University of Milan, Italy
Florbela Correia	Instituto Politécnico de Viana do Castelo, Portugal
Heliodoro Cruz-Suárez	Universidad Juárez Autónoma de Tabasco, Mexico
Patrizia Daniele	University of Catania, Italy
Mirela Danubianu	Stefan cel Mare University of Suceava, Romania
Andrea D'Ariano	Università degli Studi Roma Tre, Italy
Clarisse Dhaenens	CRIStAL, France

Brenda Dietrich	Cornell University, USA
Nikolai Dokuchaev	Curtin University, Australia
Ali Emrouznejad	Aston University, UK
Kurt Engemann	Iona College, USA
Nesim Erkip	Bilkent University, Turkey
Giovanni Fasano	University of Venezia Ca' Foscari, Italy
Luis Miguel Ferreira	Universidade de Coimbra, Portugal
Paola Festa	University of Napoli, Italy
Gary Gaukler	Drucker School of Management, USA
Claudio Gentile	Institute for System Analysis and Computer Science A. Ruberti, CNR-IASI, Italy
Ronald Giachetti	Naval Postgraduate School, USA
Stefano Giordani	University of Roma Tor Vergata, Italy
Alessandro Giuliani	University of Cagliari, Italy
Giorgio Gnecco	IMT - University of Lucca, Italy
Marc Goerigk	Lancaster University, UK
Boris Goldengorin	Ohio University, USA
Marta Gomes	CERIS-CESUR, Instituto Superior Técnico, Universidade de Lisboa, Portugal
Dries Goossens	Ghent University, Belgium
Stefano Gualandi	University of Pavia, Italy
Christelle Guéret	University of Angers, France
Francesca Guerriero	University of Calabria, Italy
David Gurzick	Hood College, USA
Gregory Gutin	Royal Holloway University of London, UK
Jin-Kao Hao	University of Angers, France
Kenji Hatano	Doshisha University, Japan
Emmanuel Hebrard	LAAS-CNRS, France
Hanno Hildmann	TNO, The Netherlands
Sin C. Ho	The Chinese University of Hong Kong, Hong Kong, China
Han Hoogeveen	Universiteit Utrecht, The Netherlands
Chenyi Hu	The University of Central Arkansas, USA
Johann Hurink	University of Twente, The Netherlands
Manuel Iori	University of Modena and Reggio Emilia, Italy
Maria Isabel Gomes	UNL, Portugal
Josef Jablonsky	University of Economics, Czech Republic
Antonio Jiménez-Martín	Universidad Politécnica de Madrid, Spain
Daniel Karapetyan	University of Essex, UK
Ahmed Kheiri	Lancaster University, UK
Jesuk Ko	Gwangju University, South Korea
Gary Kochenberger	University of Colorado at Denver, USA
Michal Koháni	University of Zilina, Slovakia
Leszek Koszalka	Wroclaw University of Technology, Poland
Yong-Hong Kuo	The University of Hong Kong, Hong Kong, China
Wasakorn Laesanklang	Mahidol University, Thailand

Andre Rossi	Université Paris-Dauphine, France
Mohamed Saleh	Cairo University, Egypt
Guzmán Santafé Rodrigo	Universidad Pública de Navarra, Spain
Cem Saydam	University of North Carolina Charlotte, USA
Joachim Schauer	University of Graz, Austria
Andrea Scozzari	Universita' degli Studi Niccolo' Cusano, Italy
Laura Scrimali	University of Catania, Italy
Stefano Sebastio	Inria, France
René Séguin	Defence Research Development Canada, Canada
Patrick Siarry	University Paris 12, LiSSi, France
Roman Slowinski	Poznan University of Technology, Poland
Stefano Smriglio	University of L'Aquila, Italy
Giuseppe Stecca	Institute of Systems Analysis and Computer Science, CNR-IASI, Italy
Thomas Stützle	Université Libre de Bruxelles, Belgium
Wai Yuen Szeto	The University of Hong Kong, Hong Kong, China
Tatiana Tambouratzis	University of Piraeus, Greece
Alberto Tonda	INRA, France
Chefi Triki	University of Salento, Italy
Michael Tschuggnall	University of Innsbruck, Austria
J. M. van den Akker	Utrecht University, The Netherlands
Inneke Van Nieuwenhuyse	KU Leuven, Belgium
Paolo Ventura	Consiglio Nazionale delle Ricerche, Italy
Maria Vlasiou	Eindhoven University of Technology, The Netherlands
Santoso Wibowo	CQUniversity, Australia
Gerhard Woeginger	RWTH Aachen, Germany
Yiqiang Zhao	Carleton University, Canada
Paola Zuddas	University of Cagliari, Italy

Additional Reviewer

| Kaushik Das Sharma | University of Calcutta, India |

Invited Speakers

Federico Della Croce	Politecnico di Torino, Italy
Helena Ramalhinho Lourenço	Universitat Pompeu Fabra, Spain
Ronald Giachetti	Naval Postgraduate School, USA

Contents

Methodologies and Technologies

Value of Image-based Yield Prediction: Multi-location Newsvendor Analysis

Kannapha Amaruchkul$^{(\boxtimes)}$

Logistics Management Program, Graduate School of Applied Statistics,
National Institute of Development Administration (NIDA),
118 Serithai Road, Bangkapi, Bangkok 10240, Thailand
kamaruchkul@gmail.com

Abstract. Consider an agricultural processing company, which wants
to pre-purchase crop from different locations before a harvesting season
in order to maximize the total expected profit from all outputs subject to
multiple resource constraints. The yields for different outputs are random
and depend on the location. By remotely sensing from satellites or locally
sensing from unmanned aerial vehicle, the firm may employ an image-
based yield prediction model at the pre-purchase time. The distribution
of the yield differs by a location. With the sensed data, the company
updates the distribution of the yield using a regression model, whose
explanatory variable is a vegetation index from image processing. At a
more favorable location, the distribution of the yield is stochastically
larger. The objective of this paper is to quantify the added value of
image sensing in predicting crop yield. Specifically, the posterior yield
distribution from image processing is used as an input to the multi-
location newsvendor model with random yields. The optimal expected
profit given the posterior distribution is compared to that with only the
prior distribution of the yield. The difference between the total expected
profits with the prior and posterior distributions is defined as the value of
the sample information. We derive the type-1 and -2 errors as a function
of the standard error of the estimate. In the numerical example, we show
that the value of the sample information tends to be increasing (with
diminishing return) as the yield prediction model becomes more accurate.

Keywords: Image processing · Agricultural supply chain · Applied
operations research · Stochastic model applications

1 Introduction

Uncertainty greatly affects the yield of agricultural produce. Agricultural yields
are affected by both exogenous and endogenous variables such as weather and
climate conditions, soil quality, irrigation techniques, pest control actions, fertil-
ization programs, and an efficiency of a production facility. Precision agriculture
(PA) attempts to optimize and improve agricultural processes, by providing fast
and reliable measurements of ongoing detailed situation in cultivation areas and

© Springer Nature Switzerland AG 2020
G. H. Parlier et al. (Eds.): ICORES 2019, CCIS 1162, pp. 3–22, 2020.
https://doi.org/10.1007/978-3-030-37584-3_1

through coordinating the automated machinery [33]. Yield prediction, one of the most important topics in PA, applies predictive analytics techniques to the "big data" obtained from sensors, both locally and remotely. Localized sensing is carried out at agricultural fields, examples are wired or wireless sensor networks at greenhouses, underground wireless sensing and image sensing by unmanned aerial vehicle (UAV). Examples of remote sensing are microwave remote sensing and optical remote sensing from satellites. From image processing, we obtain various vegetation indices (VIs) and canopy biophysical parameters, which are correlated to biomass and yield. Nevertheless, how they improve yield prediction and their added value to an agricultural processing firm are still unclear. In this article, we want to quantify the monetary benefit of image processing for predicting crop yield to an agricultural processing company.

The overview of remote sensing-based methods for predicting crop yield can be found in [20] and [23]. The remote sensing systems used for mapping crop areas and forecasting crop yield reflective (optical), thermal and microwave, whose main advantage is its ability of acquiring images under any weather conditions, such as cloud cover, rain, snow and solar irradiance. Measurements of the electromagnetic spectrum can be made from satellite, ground-based systems, aircraft or UAV. [26] evaluates the potential for drone images to assess the degree of canopy closure in order to predict yield in sugarcane fields. Spectral VIs from image processing are used along with other explanatory variables such as crop transpiration in [5], climate and weather data in [4] and [16]. The methodologies developed for yield prediction can be broadly classified as data-driven model and mechanistic crop growth models. [17] surveys advanced analytics for big data sensing in agriculture. With the exception of [16], these articles focus on assessing the technologies that would improve the predictive model. However, they do not focus on the model accuracy with inclusion of sensing data. In this article, we provide a framework to quantify the value of information for an agricultural processing company, which may use images to predict crop yields, in order to determine the pre-purchase quantities from different cultivation areas.

A framework to assess the value of information and cost of privacy in IoT is presented in [31]. If a business model is perceived to be beneficial by both customers and providers, then technological innovations would often be successful. In an agricultural supply chain, both farmers and agricultural processing firms can benefit from an accurate yield prediction model. One of the objectives of this paper is to understand how the image-sensing data helps improving the forecast accuracy. In particular, we identify the benefit of advanced information to the agricultural processing firm, which consider whether or not to pre-purchase crop, before harvest season, from certain fields. [28] quantifies the value of advanced information in minimizing the effects of supply chain disruptions. [25] quantifies the value of flexible backup suppliers and disruption risk information using a newsvendor analysis with recourse.

In a newsvendor (single-period inventory) model, a decision maker has a single opportunity to place an order before knowing an actual demand. Leftovers cannot be kept from one selling season to the next due to perishability.

The newsvendor analysis has been applied to style goods and perishable items such as bakery and agricultural produce. A literature review of the newsvendor model can be found in [24] and [8]. The multi-item newsvendor model was first formulated in [11]: The model determines optimal order quantities for all items so that the total expected profit is maximized subject to a single resource constraint. When there are few items, the exact solution is obtainable using a dynamic programming approach. When the number of items is large, several heuristics are proposed in, e.g., [22]. [19] extends the multi-item newsvendor analysis to include a fixed ordering cost. [15] extends the multi-item newsvendor problem to include multiple resource constraints. The multi-product multi-constraint newsvendor model is a convex optimization problem. Heuristic solutions based on a quadratic programming approach are presented in, e.g., [1] and [6]. The multi-product newsvendor problem is reviewed in [32] and [7]. Most of these models assume one supplier at one location [32]. [12] presents the multi-location newsvendor model, in which the demands across store locations are strongly correlated and the decision makers' psychological disutility for leftovers is stronger than that for stockouts. [10] develops a multi-location newsvendor model, which determines production quantities with respect to the privacy of local information.

In the multi-product newsvendor model, an end-item demand of a product depends on a single input. In manufacturing/production setting, an end-item demand of a product depends on multiple inputs, and the bill of material (BOM) specifies the quantities of raw materials, subassemblies, parts and components to manufacture a product. In agricultural supply chain, after an input is processed, many outputs are obtained. For instance, in a rice processing industry, outputs from milling paddies include head rice (the highest value product), rice bran, husk and broken rice [34]. The percentage of head rice is large, if a high quality paddy is processed. The head rice yield is random, and the distribution of the yield is specified by the input type. [3] presents a multi-input and -output newsvendor model with random yield. We extent [3] to include multiple resource constraints: An agricultural processing firm wants to pre-purchase rice paddies from different field locations subject to several resource constraints. The distribution of the yield depends on the field location. We can think of different locations as different input types. The objective function (the total expected profit from all outputs) is similar to [3], but in this article the resource constraints are explicitly accounted for. Furthermore, in ours, the agricultural firm may acquire additional information regarding the distribution of the yield by employing an image-based yield prediction model. In our paper, the posterior distribution is constructed from image processing, whereas in [3] the firm possesses only the prior distribution, and no further information is available. Specifically, we compare the expected profit without additional information and that with the image sensing data, and the profit gain is defined as the value of sample information.

Operations research (OR) models applied to the agriculture supply chain are reviewed in [14,29] and [2]. The demand uncertainty and the random yield are two of the most important characteristics of the agriculture OR models, and they are presented in [30] and [13]. [30] studies a planning problem, in which the production quantity at each farm is random, depending on the planted area and the crop yield in that farm. [13] formulates a two-stage stochastic programming problem for the production planning in the olive oil industry; the oil producer can buy olives from farmers at a unit cost varying with the yield. Like ours, [30] includes multiple farm fields, but we allow the decision maker to update the distribution of yields at different fields using an image-based VI, and we quantify the value of advanced information associated with the image-based yield prediction model. We include all outputs after an input is processed, whereas [13] considers a single output from a single input. In [13], the decision maker can place an additional order to an external supplier, after the actual yield is realized. In ours, the decision maker determines pre-purchase quantities based on the distribution of the yields at different locations, not the actual yield. The posterior yield distribution can be updated from an image-based VI.

The rest of this paper is organized as follows: Sect. 2 presents the multi-location newsvendor model with random yield. In Sect. 3 we define the value of perfect information, if we had known the actual yield. The value of perfect information is the maximum amount we are willing to pay for the yield prediction model. In Sect. 4, we define the value of sample information associated with the image-based yield prediction model. We illustrate the application of our approach to the rice processing industry and provide some managerial insights in Sect. 5. Finally, the conclusion is provided in Sect. 6.

2   Multi-location Newsvendor with Random Yields in Agriculture

Consider an agricultural firm that faces random demands of n different outputs, $\mathbf{D} = (D_1, D_2, \ldots, D_n)$. Let p_j be the per-unit selling price of a type-j output (given in THB/ton). Let the type-1 output denote the main product; i.e., $p_1 \gg p_j$ for all other byproducts $j > 1$. Before a harvest season, the firm wants to pre-purchase crop from m potential farm fields. Let x_i be the crop area (given in hectare (ha)) to be contracted from field i, and c_i the wholesale cost (given in THB/ha) at field i. Although the wholesale price may be smaller than the actual crop price during the harvest season, the farmer does not bear any risks associated with yield uncertainty and/or price volatility of agricultural products: The agricultural firm pre-purchases at a cheaper price but needs to hedge against the uncertainty in demand and yield.

One of the most important factors in determining the pre-purchase quantity is the yield of the outputs (given in ton/ha), associated with each location. Let $A_{ij}(\omega)$ denote the yield of a type-j output from field i, given that the state of the nature is $\omega \in \Omega$ where Ω denotes the set of all scenarios. Let $(\Omega, \mathcal{A}, P)$ be a probability space, where $\mathcal{A}$ is the set of events and P is a function from events to probabilities. The agricultural firm needs to decide $\mathbf{x} = (x_1, x_2, \ldots, x_m)$ before knowing the actual values of the yield and the demand for all outputs. The probability distribution of the demand $\mathbf{D}$ can be constructed using the historical data and the current market conditions. Similarly, the firm can empirically construct its belief regarding the state of nature, i.e., the probability space $(\Omega, \mathcal{A}, P)$, using the yields in the previous seasons. Furthermore, the firm may use some additional information from this season (say, data from image sensing at the pre-purchase time before this season's harvest) to adjust its belief regarding the yield at each location. We can employ the yield prediction model, whose inputs may include some vegetation indices calculated from images and some meteorological parameters (e.g., the surface temperature and rainfall). If the accuracy of the prediction model is high, it may worth the cost of obtaining these inputs and passing through the prediction model. The main objective of this article is to quantify the monetary value of the information from image sensing.

Let $(y)^+ = \max\{y, 0\}$, and we use E_X to denote the expectation with respect to the distribution of a random variable X. The volume of a type-j output given the input $\mathbf{x}$ is given as

$$Y_j(\mathbf{x}, \omega) = \sum_{i=1}^{m} x_i A_{ij}(\omega). \tag{1}$$

Let h_j be the per-unit salvage value of a type-j output and g_j the per-unit penalty cost for a type-j shortage (both are given in THB/ton). The type-j sales are $S_j(\mathbf{x}) = \min(Y_j(\mathbf{x}), D_j)$, the leftovers are $W_j(\mathbf{x}) = (Y_j(\mathbf{x}) - D_j)^+$, and the shortages are $T_j(\mathbf{x}) = (D_j - Y_j(\mathbf{x}))^+$. If the yield is in scenario ω, the expected profit (with respect to the demand distribution) is given as:

$$\pi(\mathbf{x}, \omega) = \sum_{j=1}^{n} E_{\mathbf{D}} \Big[p_j S_j(\mathbf{x}, \omega) + h_j W_j(\mathbf{x}, \omega) - g_j T_j(\mathbf{x}, \omega) \Big].$$

Using a Lebesque integral, we define the expected profit as:

$$\pi(\mathbf{x}) = \int_{\Omega} \pi(\mathbf{x}, \omega) P(d\omega). \tag{2}$$

We have τ resource constraints. For each $k = 1, 2, \ldots, \tau$, let $t_k \geq 0$ be the available type-k resource. One unit of a type-i input requires $\gamma_{ki} \geq 0$ units of a type-k resource. Also one unit of a type-j output requires $\eta_{kj} \geq 0$ units of a type-k resource. Let $\boldsymbol{\alpha} = (\alpha_1, \ldots, \alpha_m)$ and $\boldsymbol{\beta} = (\beta_1, \ldots, \beta_m)$ denote the lower and upper bounds on the pre-purchase quantities from m locations, respectively. Assume that we are risk neutral, and we want to maximize the expected total profit subject to the resource constraints:

$$\pi^{(1)} = \max_{\mathbf{x} \in [\boldsymbol{\alpha}, \boldsymbol{\beta}]} \left\{ \pi(\mathbf{x}) : \sum_{i=1}^{m} \gamma_{ki} x_i + \sum_{j=1}^{n} \eta_{kj} Y_j(\mathbf{x}, \omega) \leq t_k, \quad k = 1, 2, \ldots, \tau, \ \omega \in \Omega \right\} \tag{3}$$

$$= \max \left\{ \int_{\Omega} \pi(\mathbf{x}, \omega) P(d\omega) : \mathbf{x} \in \bigcap_{\omega \in \Omega} C(\omega) \right\}, \tag{4}$$

where

$$C(\omega) = \left\{ \mathbf{x} \in [\boldsymbol{\alpha}, \boldsymbol{\beta}] : \sum_{i=1}^{m} \gamma_{ki} x_i + \sum_{j=1}^{n} \eta_{kj} Y_j(\mathbf{x}, \omega) \le b_k, \quad k = 1, 2, \ldots, \tau \right\}.$$

When the agricultural firm does not have any additional information besides the prior belief on the yield $(\Omega, \mathcal{A}, P)$, the optimal expected profit $\pi^{(1)}$ is obtained. In the optimization problem (3), there are $\tau|\Omega|$ constraints. For many practical problems, the number of constraints may be much smaller. If all outputs do not require any resources (i.e., $\eta_{kj} = 0$), then there are τ linear constraints. For instance, the problem with the budget constraint t_1 has only one constraint $\sum_{i=1}^{m} c_i x_i \le t_1$. If a truck with capacity of t_2 cubic meter is used for collecting all inputs $\mathbf{x}$, and each unit of a type-i input requires w_i cubic meter, then we have the constraint $\sum_{i=1}^{m} w_i x_i \le t_2$.

Our formulation subsumes the standard multi-product newsvendor model with multiple constraints (a.k.a., the "newsstand" model in [15] and [32]). Specifically, (3) becomes the multi-product newsvendor problem, when there is one location ($m = 1$) and the yield is 100%, i.e., $Y_{1j} = A_{1j} x_j = x_j$ for each output $j = 1, \ldots, n$.

Analysis of Multi-location Newsvendor

Assume that $p_j - h_j + g_j \ge 0$ for each $j = 1, 2, \ldots, n$. This assumption is not restrictive; in most practical cases, the per-unit selling price p_j exceeds the per-unit salvage value h_j, and the assumption holds.

Theorem 1. *The optimal expected profit $\pi^{(1)}$ increases as one of the following conditions hold ceteris paribus: (i) the per-unit selling price p_j increases; (ii) the per-unit salvage value h_j increases; (iii) the per-unit penalty cost h_j decreases; (iv) the per-unit cost of input c_i decreases; (v) the distribution of the yield A_{ij} is larger in the increasing concave order; (vi) the distribution of the demand D_j is larger in the concave order (or increasing concave order if all $p_j = 0$); (vii) the resource available t_k increases; (viii) the resource requirement of input γ_{ki} decreases; (ix) the resource requirement of output η_{kj} decreases; and (x) the bounds become larger (i.e., α_i decreases or β_i increases).*

Proof. For each realization of yield A_{ij} and demand D_j, the net revenue (defined as the revenue minus the penalty cost) at location j can be written as

$$\pi_j(\mathbf{x}; A_{ij}, D_j) = p_j \min\left(\sum_{i=1}^{m} A_{ij} x_i, D_j\right) + h_j \left(\sum_{i=1}^{m} A_{ij} x_i - D_j\right)^+ - g_j \left(D_j - \sum_{i=1}^{m} A_{ij} x_i\right)^+$$

$$= (p_j - h_j + g_j) \min\left(\sum_{i=1}^{m} A_{ij} x_i, D_j\right) + h_j \sum_{i=1}^{m} A_{ij} x_i - g_j D_j.$$

Clearly, the profit increases as p_j increases, h_j increases and g_j decreases. The assumption $p_j - h_j + g_j \ge 0$ implies that $\pi_j(\mathbf{x}; A_{ij}, D_j)$ is increasing and concave in A_{ij}. Thus, the expected profit increases if A_{ij} is larger in the increasing concave order. Furthermore, $\pi_j(\mathbf{x}; A_{ij}, D_j)$ is concave in D_j, and the expected profit increases if D_j is larger in the concave order. Specifically, if $g_j = 0$ for all j, then the last term vanishes, and $\pi_j(\mathbf{x}; A_{ij}, D_j)$ is increasing concave in D_j, and the expected profit increases if D_j increases in increasing concave order. (Definitions and results on the stochastic orders can be found in [27] and [21].) Finally, the constraint set $C(\omega)$ becomes larger if one of the conditions (vii)–(x) holds, so the optimal profit increases.   □

Theorem 1 make economic sense. Intuitively, increasing the selling price (while keeping the demand relatively the same in some inelastic markets) would lead to a profit gain. Suppose that the yield is normally distributed. Then, increasing the average yield and decreasing yield variability would lead to a better profit. (Specifically, for normal random variables X and Y, $X \leq_{icv} Y$ if $E[X] \leq E[Y]$ and $\mathrm{var}(X) \geq \mathrm{var}(Y)$.) In precision agriculture, reducing yield variability may be achieved by continuously monitoring and controlling the agricultural environment. A sensitivity analysis with respect to the resource availability can be used to evaluate the benefit from acquiring additional units of resource.

Theorem 2. *The optimization problem (3) is a convex programming problem.*

Proof. The objective function is concave in $\mathbf{x}$; see [3]. Substituting $Y_j(\mathbf{x}, \omega) = \sum_{i=1}^m x_i A_{ij}(\omega)$ into the constraint (3), we obtain

$$\sum_{i=1}^m \left[\gamma_{ki} + \left(\sum_{j=1}^n \eta_{kj} A_{ij}(\omega) \right) \right] x_i \leq t_k, \quad k = 1, 2, \ldots, \tau, \ \omega \in \Omega.$$

The constraints are linear; thus the constraint set $C(\omega)$ is a convex set for each realization $\omega \in \Omega$. In (4), the intersection of convex sets is also convex. The maximization problem with a concave objective function and a convex constraint set is a convex programming problem.   □

Theorem 2 ensures that a local optimum is also global optimum. A convex optimization problem is more general than a linear programming problem, but it shares many desirable properties of linear programming such as duality and efficient algorithms. Available off-the-shelf software packages for a convex optimization includes CPLEX, CVX and MONSEK. Excel solver efficiently handles some special classes of convex programming.

3   Value of Perfect Information

To calculate the expected value of perfect information (EVPI), we suppose that we could somehow know the scenario ω. Let

$$\mathbf{x}^*(\omega) = \mathrm{argmax}\{\pi(\mathbf{x}, \omega) : \mathbf{x} \in C(\omega)\}$$

denote our optimal action associated with one particular scenario ω. Under the perfect information, the maximum profit is

$$\pi^{(0)} = \int \pi(\mathbf{x}^*(\omega), \omega) P(d\omega). \tag{5}$$

The expected value of perfect information is defined as

$$\text{EVPI} = \pi^{(0)} - \pi^{(1)}.$$

Theorem 3. $\pi^{(0)} \geq \pi^{(1)}$; thus, $EVPI \geq 0$.

Proof. Since $\mathbf{x}^*(\omega)$ is an optimal solution given that the scenario is ω, we have that

$$\pi(\mathbf{x}^*(\omega), \omega) \geq \pi(\mathbf{x}^*, \omega)$$

where

$$\mathbf{x}^* = \text{argmax} \left\{ \int \pi(\mathbf{x}, \omega) P(d\omega) : \mathbf{x} \in \bigcap_{\omega \in \Omega} C(\omega) \right\}.$$

Taking the expectation on both sides, we get

$$\int \pi(\mathbf{x}^*(\omega), \omega) P(d\omega) \geq \int \pi(\mathbf{x}^*, \omega) P(d\omega).$$

With the perfect information, the feasible region is $C(\omega)$, whereas without any information the feasible region is $\cap_{\omega \in \Omega} C(\omega)$. The feasible region in (4) is smaller than that in the problem with the perfect information (5); hence, $\pi^{(0)} \geq \pi^{(1)}$. $\square$

The EVPI is the maximum we are willing to pay for yield prediction. In our numerical example, we will see that if yield variability is larger, we are willing to pay more for the information. The cost of the image-based prediction system may include the costs of maintenance and services, installation costs of hardware and localized commercial sensors, image acquisition costs, costs of data transfer from devices to the data center and costs of big data analytics solutions.

Scenario Representation of Discretized Yield

Assume that $A_{i1} = a_i \xi$ where a_i is the deterministic component at a processing facility at location i and ξ is the random component. The output yield depends on both the efficiency of the processing facility and the random yield at the field. The uncertainty at the field is much greater than that at the processing facility, so we assume that a_i is deterministic and ξ is stochastic. For instance, in the rice processing industry, the head rice yield of a paddy field at location i can be written as $A_{i1} = (1 - \delta_i)\xi$ where ξ is the rice yield (given in ton of paddies per one hectare) and $\delta_i \in (0, 1]$ is the percent loss due to the inefficiency of the milling process and the percent moisture content of the paddy from location i.

In the precision agriculture in which an environment at the processing facility is monitored real-time, the deterministic parameter a_i may be estimated from the sensor data. For the output types of the byproducts $(j > 1)$, we assume that in each scenario $\omega_k \in \Omega$, $A_{ij}(\omega_k) = q_j A_{i1}(\omega_k)$ where q_j is the number of units (deterministic quantity) of the type-j output per one unit of the main product after processing. This assumption holds for many crops and allows us to focus on the uncertainty of the yield of the main output.

Associated with a continuous random variable X, we let f_X denote the density function, F_X the cumulative distribution function (CDF) $\bar{F}_X$ the complementary cumulative distribution function (CCDF), and $F_\xi^{-1}(\alpha)$ the α quantile. To avoid technical difficulty from the uncountable set, we assume that there are κ states of nature. (This is not restrictive in practice since κ can be a very larger positive integer.) We denote the state of nature by the random variable S, i.e., a measurable function from the probability space $(\Omega, \mathcal{A}, P)$ to the finite set with κ elements. Using the set of nondecreasing breakpoints $\{b_k^s : k = 0, 1, 2, \ldots, \kappa\}$, we discretize the yield ξ into κ bins: The state of nature is

$$S = \omega_k \text{ if } \xi \in (b_{k-1}^s, b_k^s]. \tag{6}$$

Given that the state of nature is ω_k, the yield can be taken as the "midpoint" in $(b_{k-1}^s, b_k^s]$, e.g., the expected conditional expectation $E_\xi[\xi | b_{k-1}^s < \xi \le b_k^s]$ or the quantile $F_\xi^{-1}((q_{k-1}^s + q_k^s)/2)$ where $q_k^s = F_\xi(b_k^s)$. The scenario representation of the yield distribution allows us to significant reduce the computational effort. With the scenario representation, the expected profit in the case of perfect information can be written as

$$\pi^{(0)} = \sum_{k=1}^{\kappa} \pi(\mathbf{x}^*(\omega_k), \omega_k) P(S = \omega_k),$$

which involves a finite summation instead of the integration in (5).

4 Value of Sample Information: Predicting Yield Using Regression

The perfect information regarding the true state of nature is rarely achieved in practice. Our information source is often subject to considerable error. Yield prediction models can be classified into two groups, namely the mechanistic growth model and the data-driven model. In the first, crop growth and yield formation processes are dynamically simulated, whereas in the latter, the yield is related to various predictors based on empirical relationships derived from measures or observed historical data: The model structure can be expressed as

$$\text{Yield} = f(\text{vegetation indices, climate, soil parameters, irrigation level, } \ldots) + \epsilon \tag{7}$$

where ϵ is the random error. Assume that ϵ is normally distributed with the mean of zero and with the constant variance. Let σ_e denote the standard deviation of the error (the standard error of the estimate (SEE)). Statistical techniques (e.g., multiple regression, polynomial regression and spline regression) and machine learning algorithms (e.g., neural network, and random forest) can be employed to find the fitting function $\hat{f}$. The model structure (7) is suitable for a particular region and a particular time; additional analysis might be needed to account for spatiotemporal variability in yield. Soil parameters such as soil moisture, soil temperature, humidity can be obtained from underground wireless sensors. Vegetation indices (VIs) are computed from images, whose sources can be either remote-sensing satellite images or localized sensing from camera sensors or UAV.

A VI is a spectral transformation that accentuates the properties of green plants so that they appear distinct from other image features. Common VIs are normalized difference vegetation index (NDVI), ratio vegetation index (RVI), enhanced vegetation index (EVI), soil-adjusted vegetation index (SAVI), different vegetation index (DVI) and perpendicular vegetation index (PVI). The VIs can be used to distinguish between soil and vegetation and to indicate the amount of vegetation (e.g., biomass and yield). Other parameters used in remote sensing include the leaf area index (LAI) and the fraction of photosynthetically active radiation (fPAR). In this article, we focus on the NDVI, since it is commonly used in previous studies [20].

The NDVI is calculated as

$$NDVI = \frac{NIR - R}{NIR + R} \tag{8}$$

where R and NIR are the surface reflectance values of red and near infrared, respectively. From (8), the NDVI value for a given pixel ranges from -1 to 1. The value close to zero indicates no green leaves, where the value close to one indicates very dense vegetative canopy. NDVI is often the index of choice in many yield prediction models; see, e.g., [18] and [20]. Under the simple linear regression model, the yield is assumed to be

$$\xi = \beta_0 + \beta_1 \eta + \epsilon \tag{9}$$

where η is the NDVI and ϵ is the error. The intercept and slope parameters (β_0, β_1) are estimated from the historical data. The higher NDVI value indicates "healthier" vegetation: The slope $\beta_1 > 0$ measures the increase in yield as the NDVI increases by one unit. With the scenario representation, we discretize the values of η into κ bins. We say that the survey result is

$$R = r_k \text{ if } \eta \in (b_{k-1}^r, b_k^r] \tag{10}$$

where $\{b_k^r : k = 1, 2, \ldots, \kappa\}$ is a nondecreasing sequence of breakpoints.

Theorem 4. *The prior probability of the state of nature is*

$$P(S = \omega_k) = F_\xi(b_k^s) - F_\xi(b_{k-1}^s). \tag{11}$$

If the survey result is $R = r_t$, the posterior probability of the state of nature is given as in the Bayes' theorem:

$$P(S = \omega_k | R = r_t) = \frac{P(R = r_t | S = \omega_k) P(S = \omega_k)}{\sum_{k=1}^{\kappa} P(R = r_t | S = \omega_k) P(S = \omega_k)} \tag{12}$$

where

$$P(R = r_t | S = \omega_k) = \frac{\int_{b_{k-1}^s}^{b_k^s} [\Phi((\omega - \beta_0 - \beta_1 b_k^r)/\sigma_e) - \Phi((\omega - \beta_0 - \beta_1 b_{k-1}^r)/\sigma_e)] f_\xi(\omega) d\omega}{F_\xi(b_k^s) - F_\xi(b_{k-1}^s)}. \tag{13}$$

Proof. The prior distribution (11) follows immediately from the construction in (6). Using (9) and (10), we have that

$$P(R = r_r, S = \omega_k) = P\Big((\xi - \epsilon - \beta_0)/\beta_1 \in (b_{k-1}^r, b_k^r], \; \xi \in (b_{k-1}^s, b_k^s]\Big)$$

$$= \int_{b_{k-1}^s}^{b_k^s} P\Big(b_{k-1}^r < (\omega - \epsilon - \beta_0)/\beta_1 \leq b_k^r \Big) f_\xi(\omega) d\omega$$

$$= \int_{b_{k-1}^s}^{b_k^s} [F_\epsilon(\omega - \beta_0 - \beta_1 b_k^r) - F_\epsilon(\omega - \beta_0 - \beta_1 b_{k-1}^r)] f_\xi(\omega) d\omega$$

Finally, we use

$$P(R = r_t | S = \omega_k) = \frac{P(R = r_t, S = \omega_k)}{P(S = \omega_k)}$$

to obtain (13).   □

Theorem 4 allows us to use the Bayes' theorem to update the belief on the state of nature after we know the NDVI from image processing.

Suppose that the survey result is r_t. Then, we maximize the conditional expected profit given the survey result r_t.

$$\pi^*(r_t) = \max \Big\{ \sum_{k=1}^{\kappa} \pi(\mathbf{x}, \omega_k) P(S = \omega_k | R = r_t) : \mathbf{x} \in \bigcap_{k=1}^{\kappa} C(\omega_k) \Big\}.$$

With the survey, the expected profit is

$$\pi^{(2)} = \sum_{t=1}^{\kappa} \pi^*(r_t) P(R = r_t)$$

where

$$P(R = r_t) = \sum_{k=1}^{\kappa} P(R = r_t | S = \omega_k) P(S = \omega_k).$$

The expected value of sample information (EVSI) (also, called the expected value of imperfect information (EVII)) is defined as

$$\text{EVSI} = \pi^{(2)} - \pi^{(1)}.$$

Analysis for Two States of Nature

In the simplest example with only two scenarios, we can think of "good" or "bad" scenarios when a yield is above or below average. For the rest of this section, we assume that there are two possible states of nature: $S = \{G, B\}$ where G and B denote good and bad yields and that the set of possible survey results is $R = \{F, U\}$ where F and U denote favorable and unfavorable, respectively. Let the breakpoint for the state of nature be b_1^s and let b_1^r be the breakpoint for the survey result. The type-1 error is the conditional probability that the survey result is unfavorable, given that the true state of nature is good:

$$P(U|G) = \frac{P(R = U, S = G)}{P(S = G)}.$$

The type-2 error is the conditional probability that the survey result is favorable, given that the true state of nature is bad:

$$P(F|B) = \frac{P(R = F, S = B)}{P(S = B)}.$$

Theorem 5. *The prior probability of the state of nature is*

$$P(S = B) = F_\xi(b_1^s) = 1 - P(S = G).$$

The type-1 error $P(U|G)$ is

$$\tau_1(\sigma_e|b_1^r) = \frac{\int_{b_1^s}^\infty \bar{\Phi}((t - \beta_1 b_1^r - \beta_0)/\sigma_e)f_\xi(t)dt}{\bar{F}_\xi(b_1^s)}.$$

The type-2 error $P(F|B)$ is

$$\tau_2(\sigma_e|b_1^r) = \frac{\int_{-\infty}^{b_1^s} \Phi((t - \beta_1 b_1^r - \beta_0)/\sigma_e)f_\xi(t)dt}{F_\xi(b_1^s)}.$$

For $b_1^s > \beta_0 + \beta_1 b_1^r$, $\tau_2(\sigma_e|b_1^r)$ is increasing in σ_e, and $\lim_{\sigma_e \to 0} \tau_2(\sigma_e|b_1^r) = 0$. On the other hand, for $b_1^s \leq \beta_0 + \beta_1 b_1^r$, $\tau_1(\sigma_e|b_1^r)$ is increasing in σ_e, and $\lim_{\sigma_e \to 0} \tau_1(\sigma_e|b_1^r) = 0$. In particular, if $b_1^s = \beta_0 + \beta_1 b_1^r$, then $\tau_1(\sigma_e|(b_1^s - \beta_0)/\beta_1) = \tau_2(\sigma_e|(b_1^s - \beta_0)/\beta_1)$; furthermore, as the SEE decreases to zero, the two types of errors decrease to zero, and EVSI = EVPI (i.e., $\pi^{(2)} \to \pi^{(0)}$).

Proof. The prior probability of the state of nature follows directly from (11), and the expressions for the two types of error from (13) in Theorem 4.
Consider $b_1^s \leq \beta_0 + \beta_1 b_1^r$: For $t \geq b_1^s \geq \beta_0 + \beta_1 b_1^r$, $\bar{\Phi}((t - \beta_1 b_1^r - \beta_0)/\sigma_e)$ is increasing in σ_e, so is $\tau_1(\sigma_e|b_1^r)$, and

$$\lim_{\sigma_e \to 0} \bar{\Phi}((t - \beta_1 b_1^r - \beta_0)/\sigma_e) = 0$$

by the property of the CCDF $\bar{\Phi}$. The proof for the other case is similar. Finally, suppose that $b_1^s = \beta_0 + \beta_1 b_1^r$. Then, both types of errors are equal by the symmetry of the normal distribution. Note that the constraint set under the problem with the imperfect information is identical to that with no information. As the SEE decreases to zero, the two types of errors decrease to zero: $P(U|G) = 0 = P(F|B)$, and

$$
\begin{aligned}
\pi^{(2)} &= \pi^*(F)P(R = F) + \pi^*(U)P(R = U) \\
&= \pi(\mathbf{x}^*(G), G)P(S = G|R = F)P(R = F) + \pi(\mathbf{x}^*(B), B)P(S = B|R = U)P(R = U) \\
&= \pi(\mathbf{x}^*(G), G)P(S = G) + \pi(\mathbf{x}^*(B), B)P(S = B) \\
&= \pi^{(0)}.
\end{aligned}
$$

The expected profit under imperfect information is achieved at its upper bound, the expected profit under perfect information: The EVSI is exactly equal to the EVPI.   □

Unless the breakpoint is chosen such that $b_1^s = \beta_0 + \beta_1 b_1^r$, we can guarantee to decrease only one type of error as we decrease the SEE. This tradeoff in Theorem 5 is often found in a hypothesis testing problem; see a standard textbook in introductory statistics, e.g., [9].

5   Numerical Illustration

Consider an agricultural company which distributes milled rice to both domestic and export markets. A rice yield is random and depends on various factors such as soil quality of a paddy field, weather, rainfall, solar radiation, water supply, farming practice, insect pests and diseases. After the paddy is dried, a weight loss occurs, and the percent weight loss depends on the percent moisture content. The dried paddy is stored in a silo and waits for milling. The $n = 4$ outputs of the milling process are the head rice ($j = 1$) and three byproducts, namely broken rice ($j = 2$), rice bran ($j = 3$) and husk ($j = 4$). The broken rice is separated into different grades. The rice bran is sold to the rice bran oil industry. The husk can be used in paper production and biomass power generation. Assume as in [3] that for each ton of the head rice after the milling process, we obtain on average $q_2 = 0.329$ tons of broken rice, $q_3 = 1.286$ tons of rice bran and $q_4 = 0.583$ tons of rice husk.

Assume that the demand for the type-j output, D_j, is independent and normally distributed with mean μ_j and standard deviation σ_j (in ton), as shown in Table 1. The coefficient of variation of demand is sufficiently small so that the probability of negative demand is negligible. In our example, we assume that there are no penalty costs and salvage values; $h_j = g_j = 0$ for all $j = 1, \ldots, 4$. The prices are given in THB/ton.

Table 1. Selling prices, salvage values and penalty costs.

Output type	Head	Broken	Bran	Husk
(j)	1	2	3	4
Selling price (THB/ton) p_j	120000	3720	2699.7	450
Demand: Mean (ton) μ_j	94.00	64.00	53.00	24.00
Demand: Standard deviation (ton) σ_j	14.10	9.60	7.95	3.60

Before the harvesting season, the agricultural company considers pre-purchasing paddies from $m = 4$ potential sellers (farmers) at a discount whole-sale price, given in Table 2. These four rice fields are located in the same region. Assume that the head yield of farmer i under scenario ω is

$$A_{i1}(\omega) = (1 - \delta_i)Q(\omega)$$

where δ_i is the percent loss due to the transportation from the paddy field of farmer i to the rice mill, the inefficiently of the milling process, the percent moisture content of the paddy and the environment of farm i, and $Q(\omega)$ is the rice yield (ton/ha). The wholesale cost and the percent loss of the four input types are given in Table 2. The higher the percent loss, the cheaper the wholesale cost. Assume that from the historical data, the rice yield, denoted ξ, is normally distributed with the mean of $\bar{\xi} = 2.0$ and the standard deviation of $\nu = 0.4$ ton. For illustrative purpose, we discretize the yield into $\kappa = 2$ scenarios in this numerical example. Let $\Omega = \{\omega_1, \omega_2\} = \{G, B\}$. The prior distribution regarding the state of the nature is as follows:

Table 2. The wholesale costs and the percent losses of the $m = 4$ input types.

Location i	1	2	3	4
Percent loss δ_i	0.00	0.10	0.15	0.20
Cost c_i (THB/ha)	2808.00	2626.13	2415.92	2205.71

$$P(S = G) = P(\xi > \bar{\xi}), \qquad P(S = B) = P(\xi \leq \bar{\xi}).$$

The rice yields in the good and bad scenarios are the 1st and 3rd quartiles of ξ:

$$Q(G) = F_\xi^{-1}(0.75), \qquad Q(B) = F_\xi^{-1}(0.25).$$

The agricultural processing firm cannot directly observe the true state of the nature, and it optimizes the expected profit with respect to the distribution of the state of the nature. The optimal expected profit is $\pi^{(1)} = 1256796$ THB, and the optimal land to be contracted is 81.28 ha; see Table 3.

Table 3. Solutions and profits under different cases regarding information.

Information	Profit (THB)	x_1^*	x_2^*	x_3^*	x_4^*	Total area (ha)
No survey	1256796	16.8	0.0	0.0	64.5	81.28
Perfect $S = G$	1280942	19.0	0.0	0.0	50.8	69.88
Perfect $S = B$	1234599	0.0	0.0	0.0	86.8	86.80

EVPI

Under the perfect information, the maximum profit is

$$\pi^{(0)} = \pi(\mathbf{x}^*(G), G)P(S = G) + \pi(\mathbf{x}^*(B), B)P(S = B) = 1257770.50,$$

where the optimal solutions under the two scenarios $\mathbf{x}^*(\omega_i)$ are shown in Table 3. Note that if the yield is above average (S = G), then the firm pre-purchases paddies from the total area of 69.88 ha; on the other hand, if the yield is below average (S = B), then the firm pre-purchases more, and only field 4 is chosen. Note that the solution without information lies between the solutions under the two scenarios [i.e., $x_i^*(B) \le x_i^* \le x_i^*(G)$].

We also perform a sensitivity analysis with respect to changes in the mean and variability of the rice yield. Figure 1 shows the optimal expected profit and

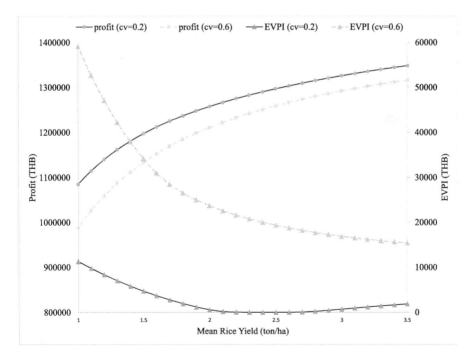

Fig. 1. EVPI and the optimal profit is affected by the mean rice yield and the yield variability.

the EVPI when the mean of the rice yield varies. The dotted (resp., solid) lines show these values when the coefficient of variation $\nu/\bar{\xi} = 0.6$ (resp., 0.2). We see that as the yield variability increases, the optimal profit decreases, but the value of information becomes larger. For instance, if the coefficient of variation of yield increases from 0.2 to 0.6, the EVPI becomes significantly much larger, from 573 to 23680 THB (given that the mean rice yield is 2.0 ton/ha). As the mean of the rice yield increases, the optimal profit increases, but the value of information becomes smaller. If the cost of the survey exceeds the EVPI, then it is not worth to obtain additional information from the survey.

EVSI: Yield Prediction Based on NDVI

For a typical rice crop, the peak/maximum greenness occurs at the end of the reproductive stage before the ripening stage, which usually takes 60–100 days after the initial seeding stage. Then, it takes around 30 days from the peak greenness stage until the harvesting stage [20]. A remotely sensed vegetation index of NDVI is low at the transplantation, increases during the reproductive stage and declines with the progression of the ripening stage. The images acquired during the peak greenness stage are commonly used in the forecasting model of rice yield. Suppose that from the historical records, the simple regression model for rice yield prediction is

$$\hat{\xi} = \beta_0 + \beta_1\eta = 0.71 + 2.24 \times \text{NDVI}, \tag{14}$$

where the SEE (standard deviation of the error term) is $\sigma_e = 0.722$ ton/ha. From (14), we interpret the slope estimate as follows: As the normalized difference vegetation index (NDVI) increases by one unit, the rice yield increases by 2.24 ton/ha. Assume that we discretize the image survey result as:

$$\text{Favorable} = F = \{\text{NDVI} > b_1^r\}$$
$$\text{Unfavorable} = U = \{\text{NDVI} \geq b_1^r\}.$$

Corollary 1. *The type-1 error is given as*

$$2\int_{b_1^r}^{\infty} \bar{\Phi}((t - \beta_1 b_1^r - \beta_0)/\sigma_e)f_\xi(t)dt,$$

and the type-2 error is given as

$$2\int_{-\infty}^{b_1^r} \Phi((t - \beta_1 b_r^1 - \beta_0)/\sigma_e)f_\xi(t)dt.$$

In particular, if b_1^r is chosen such that $\bar{\xi} = \beta_0 + \beta_1 b_1^r$, then the type-1 error is equal to the type-2 error.

Proof. The prior probability, the type-1 error and the type-2 error can be obtained directly from the results in Theorem 4.   □

Figure 2 shows the two types of error when we vary the SEE in $\{0.9, 0.8, 0.7, 0.6, 1/2, 1/2^2, \ldots, 1/2^8\}$. For the breakpoint of $b_1^r = 0.7$, we see that as the SEE decreases, the type-2 error (the dotted line in Fig. 2) decreases, whereas the type-1 error (the dashed line in Fig. 2) increases. For the breakpoint of $b_1^r = 0.4$, we see the opposite. Unless the breakpoint is $b_1^r = (\bar{\xi} - \beta_0)/\beta_1 = 0.576$, we cannot simultaneously decrease both types of errors. Figure 3 shows the profits under different conditions regarding the information available to the firm. Note that the profit with the survey is bounded above by the profit under the perfect information and the profit without the survey. Figure 3 can be used to justify the cost of getting more samples so that the SEE becomes smaller. Suppose that the breakpoint of $b_1^r = 0.7$ is chosen. To see a significant gain in the profit, the number of additional samples must be larger enough so that the SEE becomes smaller than 0.3. Furthermore, if our SEE is near 0.1, we may not need additional samples, since the further reduction in the SEE does not lead to a significant increase in the profit; see the solid line with the rectangle marker in Fig. 3. We can also decrease the SEE by allowing a nonlinear or polynomial regression.

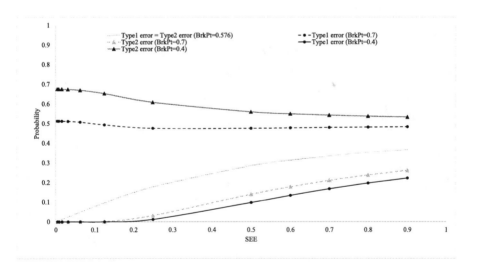

Fig. 2. Errors are affected by the SEE and the choice of breakpoint.

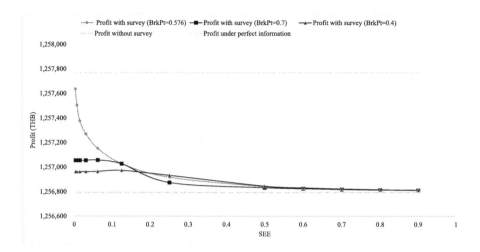

Fig. 3. Profits with sample information are affected by the SEE and the choice of breakpoint.

6   Concluding Remark

In summary, we quantify the value of information associated with the image-based yield prediction model. We provide an approach to determine the monetary gain from the increase accuracy of the prediction model. With the advanced information on the distribution of yield, the agricultural processing firm can make better decision on how to pre-purchase crop from different fields. The difference between the optimal profit with the sensing data and that without any additional data is defined as the value of sample information. The expected profit gain, or the value of sample information, can be used to justify the cost associated to the prediction model.

In our paper, we focus on the value of information from image sensing. Our approach can be extended to quantify the value of other sensors, which may continuously monitor the crop condition. We can consider contract farming, one of the most common arrangements between a farmer and a private agricultural processing company, wherein the farmer agrees to produce at a pre-agreed market price for procurement by the other party. A mechanism design framework can be applied in order to improve efficiency of the agriculture supply chain through an optimal contract, where the field condition is continuously monitored by the processing company. We hope to pursue these or related issues in the future.

Acknowledgments. The problem was materialized after some discussions with Mr. Chatbodin Sritrakul, our part-time master student who owns a rice mill in the Northeast of Thailand. His independent project, a part of requirement for a master's degree in logistics management at the school, was related to our model.

References

1. Abdel-Malek, L., Areeratchakul, N.: A quadratic programming approach to the multi-product newsvendor problem with side constraints. Eur. J. Oper. Res. **176**, 1607–1619 (2007)
2. Ahumada, O., Villalobos, J.: Application of planning models in the supply chain of agricultural products: a review. Eur. J. Oper. Res. **196**(1), 1–20 (2009)
3. Amaruchkul, K.: Newsvendor model for multi-inputs and -outputs with random yield: applciations to agricultural processing industries. In: Proceedings of the 8th International Conference on Operations Research and Enterprise Systems (ICORES 2019), Prague, Czech Republic, January 2019 (2019)
4. Cai, X., Sharma, B.: Integrating remote sensing, census and weather data for an assessment of rice yield, water consumption and water productivity in the Indo-Gangetic river basin. Agric. Water Manag. **97**, 309–316 (2010)
5. Campos, I., Neale, C., Arkebauer, T., Suyker, A.E., Goncalves, I.: Water productivity and crop yield: a simplified remote sensing driven operational approach. Agric. For. Meteorol. **249**, 501–511 (2018)
6. Chernonog, T., Goldberg, N.: On the multi-product newsvendor with bounded demand distributions. Int. J. Prod. Econ. **203**, 38–47 (2018)
7. Choi, S.: A multi-item risk-averse newsvendor with law invariant coherent mueasures of risk. In: Choi, T. (ed.) Handbook of Newsvendor Problems, vol. 176. Springer, New York (2012). https://doi.org/10.1007/978-1-4614-3600-3_2
8. Choi, T. (ed.): Handbook of Newsvendor Problems. Springer, New York (2012). https://doi.org/10.1007/978-1-4614-3600-3
9. DeGroot, M., Schervish, M.: Probability and Statistics. Addison-Wesley, Boston (2002)
10. Dobhan, A., Oberlaender, M.: Hybrid contracting within multi-location networks. Int. J. Prod. Econ. **143**, 612–619 (2013)
11. Hadley, G., Whitin, T.: Analysis of Inventory Systems. Prentice-Hall, Upper Saddle River (1963)
12. Ho, T., Lim, N., Cui, T.: Reference dependence in multilocation newsvendor models: a structural analysis. Manag. Sci. **56**(11), 1891–1910 (2010)
13. Kazaz, B.: Production planning under yield and demand uncertainty with yield-dependent cost and price. Manuf. Serv. Oper. Manag. **6**(3), 209–224 (2004)
14. Kusumastuti, R., van Donk, D., Teunter, R.: Crop-related haresting and processing planning: a review. Int. J. Prod. Econ. **174**, 76–92 (2016)
15. Lau, H., Lau, A.: The multi-product multi-constraint newsboy problem: applications, formulation and solution. J. Oper. Manag. **13**, 153–162 (1995)
16. Li, Y., Guan, K., Yu, A., Zhao, L., Li, B., Peng, J.: Toward building a transparent statistical model for improving crop yield prediction: modeling rainfed corn in the U.S. Field Crop. Res. **234**, 55–65 (2019)
17. Liakos, K., Busato, P., Moshou, D., Pearson, S., Bochtis, D.: Machine learning in agriculture: a review. Sensors **18** (2018). https://doi.org/10.3390/s18082674
18. Lisboa, I., et al.: Prediction of sugarcane yield based on NDVI and concentration of leaf-tissue nutrients in fields managed with straw removal. Agronomy **8** (2018). https://doi.org/10.3390/agronomy8090196
19. Moon, I., Silver, E.: The multi-item newsvendor problem with a budget consteraint and fixed ordering costs. J. Oper. Res. Soc. **51**(5), 602–608 (2000)
20. Mosleh, M., Hassan, Q., Chowdhury, E.: Application of remote sensors in mapping rice area and forecasting its production: a review. Sensors **15**, 769–791 (2015)

21. Müller, A., Stoyan, D.: Comparison Methods for Stochastic Models and Risks. Wiley, Chichester (2002)

22. Nahmias, S., Schmidt, C.: An efficient heuristic for the multi-item newsboy problem with a single constraint. Nav. Res. Logist. Q. **31**(3), 463–474 (1984)

23. Niel, T., McVicar, T.: Remote sensing of rice-based irrigated agriculture: a review. Rice CRC Technical report R1105–01/01 (2001)

24. Qin, Y., Wang, R., Vakharia, A., Chen, Y., Seref, M.: The newsvendor problem: review and directions for future research. Eur. J. Oper. Res. **213**(2), 361–374 (2011)

25. Saghafian, S., Oyen, M.: The value of flexible backup suppliers and disruption risk information: newsvendor analysis with recourse. IIE Trans. **44**(10), 834–867 (2012)

26. Sanches, G., et al.: The potential for RGB images obtained using unmanned aerial vehicle to assess and predict yield in sugarcane fields. Int. J. Remote Sens. **39**, 5402–5414 (2018)

27. Shaked, M., Shanthikumar, J.: Stochastic Orders. Springer, New York (2010). https://doi.org/10.1007/978-0-387-34675-5

28. Snyder, L., Atan, Z., Peng, P., Rong, Y., Schmitt, A., Sinsoysal, B.: OR/MS models for supply chain disruptions: a review. IIE Trans. **48**(2), 89–109 (2016)

29. Soto-Silva, W., Nadal-Roig, E., Gonzalez-Araya, M., Pla-Aragones, L.: Operational research models applied to the fresh fruit supply chain. Eur. J. Oper. Res. **251**, 345–355 (2016)

30. Tan, B., Comden, N.: Agricultural planning of annual plants under demand, maturation, harvest, and yield risk. Eur. J. Oper. Res. **220**, 539–549 (2012)

31. Turgut, D., Boloni, L.: Value of information and cost of privacy in the Internet of Things. IEEE Commun. Mag. **55**, 62–66 (2017)

32. Turken, N., Tan, Y., Vakharia, A., Wang, L., Wang, R., Yenipazarli, A.: The multi-product newsvendor problem: review, extensions, and directions for future research. In: Choi, T. (ed.) Handbook of Newsvendor Problems. International Series in Operations Research & Management Science, vol. 176, pp. 3–39. Springer, New York (2012). https://doi.org/10.1007/978-1-4614-3600-3_1

33. Tzounis, A., Katsoulas, N., Bartzanas, T.: Internet of Things in agriculture, recent advances and future challenges. Biosyst. Eng. **164**, 31–48 (2017)

34. Wilasinee, S., Imran, A., Athapol, N.: Optimization of rice supply chain in Thailand: a case study of two rice mills. In: Sumi, A., Fukushi, K., Honda, R., Hassan, K. (eds.) Sustainability in Food and Water: An Asian perspective, vol. 18, pp. 263–280. Springer, Heidelberg (2010). https://doi.org/10.1007/978-90-481-9914-3_27

Non-overlapping Sequence-Dependent Setup Scheduling with Dedicated Tasks

Marek Vlk[1,2(✉)], Antonin Novak[2,3], Zdenek Hanzalek[2], and Arnaud Malapert[4]

[1] Department of Theoretical Computer Science and Mathematical Logic,
Faculty of Mathematics and Physics, Charles University, Prague, Czech Republic
[2] Czech Institute of Informatics, Robotics, and Cybernetics,
Czech Technical University in Prague, Prague, Czech Republic
{marek.vlk,antonin.novak,zdenek.hanzalek}@cvut.cz
[3] Department of Control Engineering, Faculty of Electrical Engineering,
Czech Technical University in Prague, Prague, Czech Republic
[4] Université Côte d'Azur, I3S, CNRS, Nice, France
arnaud.malapert@unice.fr

Abstract. The paper deals with a parallel machines scheduling problem with dedicated tasks with sequence-dependent setup times that are subject to the non-overlapping constraint. This problem emerges in the productions where only one machine setter is available on the shop floor. We consider that setups are performed by a single person who cannot serve more than one machine at the same moment, i.e., the setups must not overlap in time. We show that the problem remains $\mathcal{NP}$-hard under the fixed sequence of tasks on each machine. To solve the problem, we propose an Integer Linear Programming formulation, five Constraint Programming models, and a hybrid heuristic algorithm LOFAS that leverages the strength of Integer Linear Programming for the Traveling Salesperson Problem (TSP) and the efficiency of Constraint Programming at sequencing problems minimizing makespan. Furthermore, we investigate the impact of the TSP solution quality on the overall objective value. The results show that LOFAS with a heuristic TSP solver achieves on average 10.5% worse objective values but it scales up to 5000 tasks with 5 machines.

Keywords: Constrained setup times · Constraint Programming · Hybrid heuristic

1 Introduction

The trend of flexible manufacturing brings many challenges typically arising from low-volume batches of a larger number of product variants. One of such challenges is the minimization of setups that must be performed when a machine switches from one product variant to another. Switching involves, e.g.,

This work was funded by Ministry of Education, Youth and Sport of the Czech Republic within the project Cluster 4.0 number CZ.02.1.01/0.0/0.0/16_026/0008432.

© Springer Nature Switzerland AG 2020
G. H. Parlier et al. (Eds.): ICORES 2019, CCIS 1162, pp. 23–46, 2020.
https://doi.org/10.1007/978-3-030-37584-3_2

tool adjustment, which requires a machine setter to reconfigure the particular machine and make the desired adjustment.

In order to cut labor costs, the companies often try to limit the number of machine setters. As a basic case, only a single machine setter is present on the shop floor to perform the reconfigurations. Consequently, any schedule that does not account for the limited capacity of a machine setter is deemed to be infeasible, as a single machine setter cannot set two machines at the same time.

In this paper, we study a scheduling problem where the tasks are dedicated to the machines and have sequence-dependent setup times. Each setup occupies an extra unary resource, i.e., the machine setter, hence, setups must not overlap in time. The goal is to minimize the makespan of the overall schedule. To solve the problem, we design an Integer Linear Programming (ILP) model, five Constraint Programming (CP) models, and heuristic algorithm LOFAS.

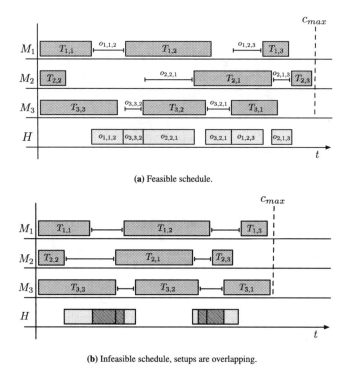

(a) Feasible schedule.

(b) Infeasible schedule, setups are overlapping.

Fig. 1. The illustration of a schedule with three machines and three tasks to be processed on each machine [17].

The same problem was addressed in [17]. This is the revised and improved version of that paper. The main contributions of this revised and improved paper with respect to [17] are:

- a complexity result for the restricted problem with fixed sequences of tasks
- two new CP models utilizing the concept of cumulative function

- LOFAS algorithm enhanced by a heuristic for the subproblem allowing to solve larger instances
- experimental results showing the impact of the subproblem solution method on the solution quality.

The rest of the paper is organized as follows. We first survey the existing work in the related area. Next, Sect. 3 gives the formal definition of the problem at hand. In Sect. 4, we describe an ILP model, while in Sect. 5 we introduce five CP models, and in Sect. 6 we propose the LOFAS algorithm. Finally, we present computational experiments in Sect. 7 and draw conclusions in Sect. 8.

2   Related Work

There is a myriad of papers on scheduling with sequence-dependent setup times or costs [2], proposing exact approaches [11] as well as various heuristics [15]. But the research on the problems where the setups require extra resource is scarce.

An unrelated parallel machine problem with machine and job sequence-dependent setup times, studied by [13], considers also the non-renewable resources that are assigned to each setup, which affects the amount of time the setup needs and which is also included in the objective function. On the other hand, how many setups may be performed at the same time is disregarded. The authors propose a Mixed Integer Programming formulation along with some static and dynamic dispatching heuristics.

A lotsizing and scheduling problem with a common setup operator is tackled in [14]. The authors give ILP formulations for what they refer to as a dynamic capacitated multi-item multi-machine one-setup-operator lotsizing problem. Indeed, the setups to be performed by the setup operator are considered to be scheduled such that they do not overlap. However, these setups are not sequence-dependent in the usual sense. The setups are associated to a product whose production is to be commenced right after the setup and thus the setup time, i.e., the processing time of the setup, does not depend on a pair of tasks but only on the succeeding task.

A complex problem that involves machines requiring setups that are to be performed by operators of different capabilities has been addressed in [5]. The authors modeled the whole problem in the time-indexed formulation and solved it by decomposing the problem into smaller subproblems using Lagrangian Relaxation and solving the subproblems using dynamic programming. A feasible solution is then composed of the solutions to the subproblems by heuristics, and, if impossible, the Lagrangian multipliers are updated using surrogate subgradient method as in [19]. The down side of this approach is that the time-indexed formulation yields a model of pseudo-polynomial size. This is not suitable for our problem as it poses large processing and setup times.

In [17], the problem with sequence-dependent non-overlapping setups is introduced. The authors propose three CP models, ILP model and heuristics utilizing a decomposition of the problem. The resulting subproblems deal with the order of tasks on machines independently. The order of tasks is found by an ILP model with lazy subtour elimination and the order of setups by a CP model.

3   Problem Statement

The problem addressed in this paper consists of a set of machines and a set of independent non-preemptive tasks, each of which is dedicated to one particular machine where it will be processed. Also, there are sequence-dependent setup times on each machine. In addition, these setups are to be performed by a human operator who is referred to as a machine setter. Such a machine settercannot perform two or more setups at the same time. It follows that the setups on all the machines must not overlap in time. Examples of a feasible and an infeasible schedule with 3 machines can be seen in Fig. 1. Even though the schedule in Fig. 1b on the machines contains setup times, such schedule is infeasible since it would require overlaps in the schedule for the machine setter.

The aim is to find a schedule that minimizes the completion time of the latest task. It is clear that the latest task is on some machine and not in the schedule of a machine settersince the completion time of the last setup is followed by at least one task on a machine.

3.1   Formal Definition

Let $M = \{M_1, ..., M_m\}$ be a set of machines and for each $M_i \in M$, let $T^{(i)} = \{T_{i,1}, ..., T_{i,n_i}\}$ be a set of tasks that are to be processed on machine M_i, and let $T = \bigcup_{M_i \in M} T^{(i)} = \{T_{1,1}, ..., T_{m,n_m}\}$ denote the set of all tasks. Each task $T_{i,j} \in T$ is specified by its processing time $p_{i,j} \in \mathbb{N}$. Let $s_{i,j} \in \mathbb{N}_0$ and $C_{i,j} \in \mathbb{N}$ be start time and completion time, respectively, of task $T_{i,j} \in T$, which are to be found. All tasks are non-preemptive, hence, $s_{i,j} + p_{i,j} = C_{i,j}$ must hold.

Each machine $M_i \in M$ performs one task at a time. Moreover, the setup times between two consecutive tasks processed on machine $M_i \in M$ are given in matrix $O^{(i)} \in \mathbb{N}^{n_i \times n_i}$, $O = \bigcup_{M_i \in M} O^{(i)}$. That is, $o_{i,j,j'} = (O^{(i)})_{j,j'}$ determines the minimal time distance between the start time of task $T_{i,j'}$ and the completion time of task $T_{i,j}$ if task $T_{i,j'}$ is to be processed on machine M_i right after task $T_{i,j}$, i.e., $s_{i,j'} - C_{i,j} \geq o_{i,j,j'}$ must hold.

Let $H = \{h_1, ..., h_\ell\}$, where $\ell = \sum_{M_i \in M} n_i - 1$, be a set of setups that are to be performed by the machine setter. Each $h_k \in H$ corresponds to the setup of a pair of tasks that are scheduled to be processed in a row on some machine. Thus, function $st : H \to M \times T \times T$ is to be found. Also, $s_k \in \mathbb{N}_0$ and $C_k \in \mathbb{N}$ are start time and completion time of setup $h_k \in H$, which are to be found. Assuming $h_k \in H$ corresponds to the setup between tasks $T_{i,j} \in T$ and $T_{i,j'} \in T$, i.e., $st(h_k) = (M_i, T_{i,j}, T_{i,j'})$, it must hold that $s_k + o_{i,j,j'} = C_k$, also $C_{i,j} \leq s_k$, and $C_k \leq s_{i,j'}$. Finally, since the machine setter may perform at most one task at any time, it must hold that, for each $h_k, h_{k'} \in H, k \neq k'$, either $C_k \leq s_{k'}$ or $C_{k'} \leq s_k$.

The objective is to find such a schedule that minimizes the makespan, i.e., the latest completion time of any task:

$$\min \max_{T_{i,j} \in T} C_{i,j} \tag{1}$$

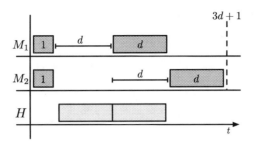

(a) An instance where optimal sequences on machines separately lead to a suboptimal solution.

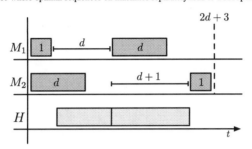

(b) Suboptimal sequence on one machine yields a globally optimal solution.

Fig. 2. Solving the problem greedily for each machine separately can lead to arbitrarily bad solutions. The numbers depict the processing times of the tasks and setups [17].

We note that minimizing the makespan for each machine separately does not guarantee globally optimal solution. In fact, such a solution can be arbitrarily bad. Consider a problem depicted in Fig. 2. It consists of two machines, M_1 and M_2, and two tasks on each machine, with processing times $p_{1,1} = p_{2,1} = 1, p_{1,2} = p_{2,2} = d$, where d is any constant greater than 2, and with setup times $o_{1,1,2} = o_{2,1,2} = d, o_{1,2,1} = o_{2,2,1} = d + 1$. Then, an optimal sequence on each machine yields a solution of makespan $3d + 1$, whereas choosing suboptimal sequence on either of the machines gives optimal objective value $2d + 3$.

In the next subsection, we study the complexity of the problem where the order of the tasks is predefined.

3.2   Complexity of the Problem with Fixed Sequences

It is easy to see that the problem with sequence-dependent non-overlapping setups is strongly $\mathcal{NP}$-hard even for the case of one machine, i.e., $m = 1$, which can be shown by the reduction from the shortest Hamiltonian path problem. However, we show that even the restricted problem where the task sequences on each machine are fixed is $\mathcal{NP}$-hard as well, suggesting another source of hardness.

Definition 1 (Problem with fixed sequences). Let us denote the position of task $T_{i,j}$ on machine M_i by $\pi_i(j)$. Then the problem with fixed sequences of tasks is defined as follows.

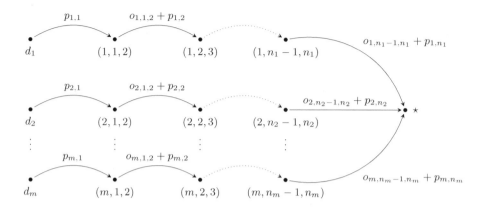

Fig. 3. Precedence graph of the problem with fixed sequences on machines.

INPUT: M, T, O, sequences of tasks for each machine $\pi_1, \pi_2, \ldots, \pi_m$.
OUTPUT: a feasible schedule with non-overlapping setups such that $\forall M_i \in M$:
$\pi_i(j) < \pi_i(j') \implies s_{i,j} < s_{i,j'}$ minimizing $\max_{T_{i,j} \in T} C_{i,j}$.

Essentially, the problem is, given a fixed sequence of tasks on each machine, find a feasible schedule for setups H such that the latest completion time of any task is minimized.

It can be seen that the problem with fixed sequences is equivalent to $1 | l_{ij} > 0, m\, n_1, \ldots, n_m\text{-chains} | C_{\max}$, i.e., a single machine scheduling problem with minimum time lags, where the precedence graph has a form of m chains of lengths $n_1, \ldots, n_m$, followed by a single common task (see Fig. 3). Indeed, consider the following reduction. Let us remind that by a *minimum time lag* $l_{j,j'} > 0$ between tasks T_j and $T_{j'}$ (i.e., $T_j \xrightarrow{l_{j,j'}} T_{j'}$) we mean that $T_{j'}$ cannot start earlier than $s_{j'} \geq s_j + l_{j,j'}$, hence $l_{j,j'}$ defines a minimum distance between the corresponding start times.

Without loss of generality, let us assume in the reduction below that the sequence of tasks on each machine $M_i \in M$ is $\pi_i = (1, 2, \ldots, n_i)$. Then, the reduction goes as follows. The sequence of setups on each machine $M_i \in M$ defines a chain of $n_i - 1$ tasks plus one dummy task d_i. The processing time of a dummy task d_i is set to $p_i = 0$. The processing times of each task representing setups are equal to the corresponding setup time $o_{i,j,j'}$ given by the tasks sequences. A minimum time lag of length $p_{i,1}$ connects each dummy task d_i to the first setup on the corresponding machine. Then, in each chain corresponding to a machine M_i, the length of time lag between j-th setup and $(j+1)$-th setup has length $o_{i,j,j+1} + p_{i,j+1}$. Finally, the last setup in each chain is connected to the dummy task $\star$ with $p_\star = 0$ by the time lag of length $o_{i,n_i-1,n_i} + p_{i,n_i}$. See an example of the graph in Fig. 3.

It is easy to see that the reduction the other way around is possible directly as well, hence establishing the problem equivalence. Finally, we note that the problem $1 | l_{ij} > 0, m\, 2\text{-chains} | C_{\max}$ is known to be strongly $\mathcal{NP}$-hard by the reduction from 3-PARTITION problem due to [18]. This suggests that the constraint requiring non-overlapping setups introduces yet another source of hardness to the classical problem with sequence-dependent setup times. We will utilize the solution for the problem with fixed sequences further in Sect. 6 as a part of the proposed heuristic algorithm.

4   Integer Linear Programming Model

The proposed model is split into two parts. The first part handles scheduling of tasks on the machines using efficient *rank-based model* [10]. This modeling uses binary variables $x_{i,j,q}$ to encode whether task $T_{i,j} \in T^{(i)}$ is scheduled at q-th position in the permutation on machine $M_i \in M$. Another variable is $\tau_{i,q}$ denoting the start time of a task that is scheduled at q-th position in the permutation on machine $M_i \in M$.

The second part of the model resolves the question, in which order and when the setups are performed by a machine setter. There, we need to schedule all setups H, where the setup time π_k of the setup $h_k \in H$ is given by the corresponding pair of tasks on the machine. The ordering of setups is determined by $z_{k,l}$ binary variables, that take value 1 if setup h_l is scheduled after the setup h_k.

Let us denote the set of all natural numbers up to n as $[n] = \{1, \ldots, n\}$. We define the following function $\phi : H \to M \times [\max_{M_i \in M} n_i]$ (e.g., $\phi(h_k) = (M_i, q)$), that maps $h_k \in H$ to setups between the tasks scheduled at positions q and $q + 1$ on machine $M_i \in M$. Since the time of such setup is a variable (i.e., it depends on the pair of consecutive tasks on M_i), the modeling with *rank-based model* would contain non-linear expressions. Therefore, we use the a disjunctive model [3,4] that admits processing times given as variables. Its disadvantage over the rank-based model is that it introduces a *big M* constant in the constraints, whereas the rank-based model does not. See Fig. 4 for the meaning of variables.

The full model is stated as:

$$\min \ c_{max}$$

$$\text{s.t.} \tag{2}$$

$$c_{max} \geq \tau_{i,n_i} + \sum_{T_{i,j} \in T^{(i)}} p_{i,j} \cdot x_{i,j,n_i} \quad \forall M_i \in M \tag{3}$$

$$\sum_{q \in [n_i]} x_{i,j,q} = 1 \quad \forall M_i \in M, \forall T_{i,j} \in T^{(i)} \tag{4}$$

$$\sum_{T_{i,j} \in T^{(i)}} x_{i,j,q} = 1 \quad \forall M_i \in M, \forall q \in [n_i] \tag{5}$$

$$s_k + \pi_k \leq s_l + \mathcal{M} \cdot (1 - z_{k,l}) \quad \forall h_l, h_k \in H : l < k \tag{6}$$

$$s_l + \pi_l \leq s_k + \mathcal{M} \cdot z_{k,l} \quad \forall h_l, h_k \in H : l < k \tag{7}$$

$$\pi_k \geq o_{i,j,j'} \cdot (x_{i,j,q} + x_{i,j',q+1} - 1)$$

$$\forall h_k \in H : \phi(h_k) = (M_i, q), \forall T_{i,j}, T_{i,j'} \in T^{(i)} \tag{8}$$

$$s_k + \pi_k \leq \tau_{i,q+1} \quad \forall h_k \in H : \phi(h_k) = (M_i, q) \tag{9}$$

$$s_k \geq \tau_{i,q} + \sum_{T_{i,j} \in T^{(i)}} p_{i,j} \cdot x_{i,j,q} \quad \forall h_k \in H : \phi(h_k) = (M_i, q) \tag{10}$$

where

$$c_{max} \in \mathbb{R}_0^+ \tag{11}$$

$$\tau_{i,q} \in \mathbb{R}_0^+ \quad \forall M_i \in M, \forall q \in [n_i] \tag{12}$$

$$s_k, \pi_k \in \mathbb{R}_0^+ \quad \forall h_k \in H \tag{13}$$

$$x_{i,j,q} \in \{0,1\} \quad \forall M_i \in M, \forall T_{i,j} \in T^{(i)}, \forall q \in [n_i] \tag{14}$$

$$z_{k,l} \in \{0,1\} \quad \forall h_k, h_l \in H : l < k \tag{15}$$

The constraint (3) computes makespan of the schedule while constraints (4)–(5) states that each task occupies exactly one position in the permutation and that each position is occupied by exactly one task. Constraints (6) and (7) guarantee that setups do not overlap. $\mathcal{M}$ is a constant that can be set as $|H| \cdot \max_i O^{(i)}$. Constraint (8) sets processing time π_k of the setup $h_k \in H$ to $o_{i,j,j'}$ if task $T_{i,j'}$ is scheduled on machine M_i right after task $T_{i,j}$. Constraints (9) and (10) are used to avoid conflicts on machines. The constraint (9) states that a task cannot start before its preceding setup finishes. Similarly, the constraint (10) states that a setup is scheduled after the corresponding task on the machine finishes.

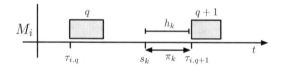

Fig. 4. Meaning of the variables in the model [17].

4.1 Additional Improvements

We use the following improvements of the model that have a positive effect on the solver performance.

1. **Warm Starts.** The solver is supplied with an initial solution. It solves a relaxed problem, where it relaxes on the condition that setups do not overlap. Such solution is obtained by solving the shortest Hamiltonian path problem given by setup time matrix $O^{(i)}$ independently for each machine $M_i \in M$. Since such solution might be infeasible for the original problem, we transform it in a polynomial time into a feasible one. It is done in the following way. Since now the permutations on machines are fixed, the problem is reduced to the problem with minimum time lags on a single machine, as described in Sect. 3.2. The solution is constructed by scheduling first m setups in the machine order, then followed by next m setups etc. Hence, for the case of fixed permutations shown in Fig. 3, the order is $(1,1,2) \prec (2,1,2) \prec \ldots \prec (m,1,2) \prec (1,2,3) \prec \ldots \prec (m, n_m - 1, n_m)$.

2. **Lower Bounds.** We supply a lower bound on c_{max} variable given as the maximum of all best proven lower bounds for corresponding shortest Hamiltonian path problem on each machine.
3. **Pruning of Variables.** We can reduce the number of variables in the model due to the structure of the problem. We fix values of some of the $z_{k,l}$ variables according to the following rule. Let $h_k, h_l \in H$ such that $\phi(h_k) = (M_i, q)$ and $\phi(h_l) = (M_i, v)$ for any $M_i \in M$. Then, $q < v \Rightarrow z_{k,l} = 1$ holds in some optimal solution. Note that the rule holds only for setups following from the same machine.

The rule states that the relative order of setups on the same machine is determined by the natural ordering of task positions on that machine. See for example setups $o_{1,1,2}$ and $o_{1,2,3}$ in Fig. 1. Since these setups follow from the same machine, their relative order is already predetermined by positions of the respective tasks. Essentially, the rule fixes the precedences according to the underlying precedence graph, as e.g., shown in Fig. 3.

5   Constraint Programming Models

A next way how the problem at hand can be tackled is by the Constraint Programming (CP) formalism, where special global constraints modeling resources and efficient filtering algorithms are used [16]. CP works with so-called *interval variables* whose start time and completion time are denoted by predicates $StartOf$ and $EndOf$, and the difference between the completion time and the start time of the interval variable can be set using predicate $LengthOf$.

We construct the CP models as follows. We introduce interval variables $I_{i,j}$ for each $T_{i,j} \in T$, and the lengths of these interval variables are set to the corresponding processing times:

$$LengthOf(I_{i,j}) = p_{i,j} \tag{16}$$

The sequence is resolved using the $NoOverlap$ constraint. The $NoOverlap(I)$ constraint on a set I of interval variables states that it constitutes a chain of non-overlapping interval variables, any interval variable in the chain being constrained to be completed before the start of the next interval variable in the chain. In addition, the $NoOverlap(I, O^{(i)})$ constraint is given a so-called *transition distance* matrix $O^{(i)}$, which expresses a minimal delay that must elapse between two successive interval variables. More precisely, if $I_{i,j}, I_{i,j'} \in I$, then $(O^{(i)})_{j,j'}$ gives a minimal allowed time difference between $StartOf(I_{j'})$ and $EndOf(I_j)$. Hence, the following constraint is imposed, $\forall M_i \in M$:

$$NoOverlap\left(\bigcup_{T_{i,j} \in T^{(i)}} \{I_{i,j}\} , O^{(i)} \right) \tag{17}$$

The objective function is to minimize the makespan:

$$\min \max_{T_{i,j} \in T} EndOf(I_{i,j}) \tag{18}$$

This model would already solve the problem if the setups were not required to be non-overlapping. In what follows we describe three ways how the non-overlapping setups are resolved. Constraints (16)–(18) are part of each of the following models.

5.1   CP1: With Implications

Let us introduce $I_{i,j}^{st}$ for each $T_{i,j} \in T$ representing the setup after task $T_{i,j}$. There is $\sum_{M_i \in M} n_i$ such variables. As the interval variable $I_{i,j}^{st}$ represents the setup after task $T_{i,j}$, we use the constraint $EndBeforeStart(I_1, I_2)$, which ensures that interval variable I_1 is completed before interval variable I_2 can start. Thus, the following constraint needs to be added, $\forall M_i \in M, \forall T_{i,j} \in T^{(i)}$:

$$EndBeforeStart(I_{i,j}, I_{i,j}^{st}) \tag{19}$$

To ensure that the setups do not overlap in time is enforced through the following constraint:

$$NoOverlap\Big(\bigcup_{T_{i,j} \in T} \{I_{i,j}^{st}\} \Big) \tag{20}$$

Notice that this constraint is unique and it is over all the interval variables representing setups on all machines. This $NoOverlap$ constraint does not need any transition distance matrix as the default values 0 are desired.

Since it is not known a priori which task will follow task $T_{i,j}$, the quadratic number of implications determining the precedences and lengths of the setups must be imposed. For this purpose, the predicate $Next^1$ is used. $Next(I)$ equals the interval variable that is to be processed in the chain right after interval variable I. Thus, the following constraints are added, $\forall M_i \in M, \forall T_{i,j}, T_{i,j'} \in T^{(i)}, j \neq j'$:

$$Next(I_{i,j}) = I_{i,j'} \Rightarrow EndOf(I_{i,j}^{st}) \leq StartOf(I_{i,j'}) \tag{21}$$

$$Next(I_{i,j}) = I_{i,j'} \Rightarrow LengthOf(I_{i,j}^{st}) = o_{i,j,j'} \tag{22}$$

Note that the special value when an interval variable is the last one in the chain is used to turn the last setup on a machine into a dummy one.

5.2   CP2: With Element Constraints

Setting the lengths of the setups can be substituted by the element constraint, which might be beneficial as global constraints are usually more efficient. More precisely, this model contains also constraints (19), (20), and (21), but constraint (22) is substituted as follows.

---

[1] Note that in the IBM CP Optimizer, the function TypeOfNext is used.

Assume the construct $Element(Array, k)$ returns the k-th element of $Array$, $(O^{(i)})_j$ is the j-th row of matrix $O^{(i)}$, and $IndexOfNext(I_{i,j})$[2] returns the index of the interval variable that is to be processed right after $I_{i,j}$. Then the following constraint is added, for each $I_{i,j}^{st}$:

$$LengthOf(I_{i,j}^{st}) = Element((O^{(i)})_j, \ IndexOfNext(I_{i,j})) \tag{23}$$

5.3   CP3: With Optional Interval Variables

In this model, we use the concept of *optional interval variables* [8]. An optional interval variable can be set to be *present* or *absent*. The predicate $PresenceOf$ is used to determine whether or not the interval variable is present in the resulting schedule. Whenever an optional interval variable is absent, all the constraints that are associated with that optional interval variable are implicitly satisfied and predicates $StartOf$, $EndOf$, and $LengthOf$ are set to 0.

Hence, we introduce optional interval variable $I_{i,j,j'}^{opt}$ for each pair of distinct tasks on the same machine, i.e., $\forall M_i \in M, \forall T_{i,j}, T_{i,j'} \in T^{(i)}, j \neq j'$. There are $\sum_{M_i \in M} n_i(n_i - 1)$ such variables. The lengths of these interval variables are set to corresponding setup times:

$$LengthOf(I_{i,j,j'}^{opt}) = o_{i,j,j'} \tag{24}$$

To ensure that the machine setter does not perform more than one task at the same time, the following constraint is added:

$$NoOverlap\Big(\bigcup_{\substack{T_{i,j},T_{i,j'} \in T \\ j \neq j'}} \{I_{i,j,j'}^{opt}\} \Big) \tag{25}$$

In this case, to ensure that the setups are indeed processed in between two consecutive tasks, we use the constraint $EndBeforeStart(I_1, I_2)$, which ensures that interval variable I_1 is completed before interval variable I_2 can start, but if either of the interval variables is absent, the constraint is implicitly satisfied. Thus, the following constraints are added, $\forall I_{i,j,j'}^{opt}$:

$$EndBeforeStart(I_{i,j}, I_{i,j,j'}^{opt}) \tag{26}$$

$$EndBeforeStart(I_{i,j,j'}^{opt}, I_{i,j'}) \tag{27}$$

Finally, in order to ensure the correct presence of optional interval variables, the predicate $PresenceOf$ is used. Thus, the following constraint is imposed, $\forall I_{i,j,j'}^{opt}$:

$$PresenceOf(I_{i,j,j'}^{opt}) \Leftrightarrow Next(I_{i,j}) = I_{i,j'} \tag{28}$$

Notice that each $I_{i,j}$ (except for the last one on a machine) is followed by exactly one setup. Thus, we tried using a special constraint called *Alternative*,

---

[2] Again, in the IBM CP Optimizer, the function TypeOfNext is used.

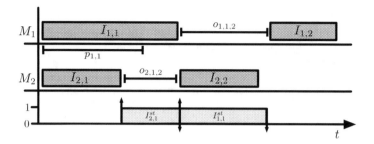

Fig. 5. Illustration of variables and constraints for CP4. The length of $I_{1,1}$ is greater than $p_{1,1}$ as the completion of $I_{1,1}$ must wait for the machine setter to complete the setup $I_{2,1}^{st}$.

which ensures that exactly one interval variable from a set of variables is present. However, preliminary experiments showed that adding this constraint is counterproductive.

5.4   CP4: With Cumulative Function

In this model, the machine setter is represented as a cumulative function to avoid the quadratic number of constraints of CP1 and CP2 or the quadratic number of optional task variables of CP3. The definition of the non-negative cumulative function has a linear number of terms.

Again, we use interval variables $I_{i,j}$ for each $T_{i,j} \in T$ to model the execution of each task $T_{i,j} \in T$ and $I_{i,j}^{st}$ for each $T_{i,j} \in T$ representing the setup after task $T_{i,j}$. See Fig. 5 for the illustration of main concepts.

The idea now is that the lengths of the interval variables are not fixed. The length of interval variables $I_{i,j}$ is increased to wait for the machine setter if he is busy on another machine, i.e., the length of interval variable $I_{i,j}$ at least the processing time task $p_{i,j}$ but may be prolonged until the machines setter is available. Thus, the constraints (16) on the lengths of the interval variables $I_{i,j}$ are relaxed because they must only be greater than the corresponding processing times:

$$LengthOf(I_{i,j}) \geq p_{i,j} \tag{29}$$

The length of the setup executed by the machine setter will be determined by the corresponding tasks. In particular, as the last setup on a machine becomes a dummy setup (of zero length), we merely set the length of the setup variables to be at least zero:

$$LengthOf(I_{i,j}^{st}) \geq 0 \tag{30}$$

The cumulative function is built thanks to primitive $Pulse(a, h)$ that specifies that h unit of resource is used during interval a. The cumulative function is composed of $Pulse$ terms for each $I_{i,j}^{st}$ representing the usage of a machine setter, and the cumulative function must remain lower than 1 because there is a single machine setter:

$$\sum_{T_{i,j} \in T} Pulse(I_{i,j}^{st}, 1) \leq 1 \tag{31}$$

What remains to be done is to synchronize start and completion times between the tasks and setups. This is done using the constraint $EndAtStart$ (I_1, I_2), which ensures that interval variable I_1 is completed exactly when interval variable I_2 starts, and by $StartOfNext(I_{i,j})$, which gives the start time of the interval variable that is to be processed right after $I_{i,j}$. Thus, the following constraints are added, for each $I_{i,j}^{st}$:

$$EndAtStart(I_{i,j}, I_{i,j}^{st}) \tag{32}$$

$$EndOf(I_{i,j}^{st}) \geq StartOfNext(I_{i,j}) \tag{33}$$

Note that the inequality in constraint (33) is necessary because $StartOfNext$ gives 0 for the last task on a machine and thus the completion time of the last setup is equal to its start time (hence the length of the setup is 0). Also, note that the setups cannot be shorter than required due to constraint (17).

5.5 CP5: Without Setup Variables

This model exploits the same idea as CP4, but we can go even further and completely omit the interval variables representing the setups. In this model, the interval variables $I_{i,j}$ are, again, relaxed because they must only be greater than the corresponding processing times, i.e., constraint (29) is kept.

The main difference in this model is that the cumulative function only requires the introduction of interval variables S_i, one for each machine M_i. See Fig. 6 for the illustration of main concepts. Variable S_i starts with the first task of the machine M_i and ends with the last task, which is enforced by the $Span$ constraint:

$$Span\left(S_i, \bigcup_{T_{i,j} \in T^{(i)}} \{I_{i,j}\}\right) \tag{34}$$

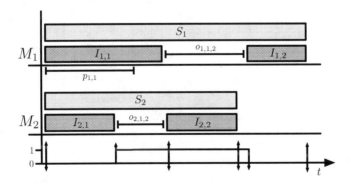

Fig. 6. Illustration of variables and constraints for CP5.

The cumulative function is now realized as follows:

$$\sum_{M_i \in M} Pulse(S_i, 1) - \sum_{T_{i,j} \in T} Pulse(I_{i,j}, 1) \leq 1 \qquad (35)$$

The first term enforces that the cumulative function remains non-negative when the first task $T_{i,j}$ of the machine M_i starts and that the machine setter is not required at the end of the last task $T_{i,j}$ of the machine M_i. The second term enforces that the machine setter is available when a task $T_{i,j}$ ends, possibly after a waiting time as allowed by constraints (29), and that the machine setter becomes available again when a task $T_{i,j}$ starts.

Despite the lowest number of variables, the main drawback of this model is that the schedule of the machine setter becomes implicit which leads to less filtering.

5.6   Additional Improvements

We use the following improvements:

1. **Search Phases.** Automatic search in the solver is well tuned-up for most types of problems, leveraging the newest knowledge pertaining to variable selection and value ordering heuristics. In our case, however, preliminary results showed that the solver struggles to find any feasible solution already for small instances. It is clear that it is easy to find some feasible solution, e.g., by setting an arbitrary order of tasks on machines and then shifting the tasks to the right such that the setups do not overlap. To make the solver find some feasible solution more quickly, we set the search phases such that the sequences on machines are resolved first, and then the sequences of setups for the machine setter are resolved. This is included in all the CP models described.

2. **Warm Starts.** Similarly to improvement (1) in Sect. 4.1, we boost the performance by providing the solver with a starting point. We do this only for CP3 as the preliminary numerical experiments showed a slight superiority of CP3.

   More precisely, we first find an optimal sequence of tasks minimizing makespan on each machine separately and then we set those interval variables $I_{i,j,j'}^{opt}$ to be present if $T_{i,j'}$ is sequenced directly after $T_{i,j}$ on machine M_i. This is all that we set as the starting point. Notice that unlike in Sect. 4.1, we do not calculate the complete solution but we let the solver do it. The solver then quickly completes the assignment of all the variables such that it gets a solution of reasonably good objective value.

   Note that the optimal sequences on machines are solved using ILP so it can be seen as a hybrid approach. This model with warm starts is in what follows referred to as *CP3ws*.

6 LOFAS Heuristics

We propose an approach that guides the solver quickly towards solutions of very good quality but cannot guarantee optimality of what is found. There are two main phases of this approach. In the first phase, the model is decomposed such that its subproblems are solved optimally or near-optimally and then the solutions of the subproblems are put together so as to make a feasible solution of the whole problem. In the second phase, the solution found is locally improved by repeatedly adjusting the solution in promising areas. More details follow.

6.1 Decomposition Phase

The idea of the model decomposition is as follows. First, we find a sequence of tasks minimizing makespan on each machine separately. Second, given these sequences on each machine, the setups to be performed are known, hence, the lengths of the setups are fixed as well as the precedence constraints with respect to the tasks on machines. This is the problem with fixed permutations described in Sect. 3.2.

The pseudocode for obtaining the initial solution is given in Algorithm 1. It takes one machine at a time and finds a sequence for it while minimizing makespan. The time limit for the computation of one sequence on a machine is given in such a way that there is a proportional remaining time limit for the rest of the algorithm. $\text{SEQ}(i, TimeLimit)$ returns the best sequence it finds on machine $M_i \in M$ in the given $TimeLimit$. The $TimeLimit$ is computed using $RemainingTime()$, which is the time limit for the entire run of the algorithm minus the time that already elapsed from the beginning of the run of the algorithm. Just in case that no sequence is found in the given $TimeLimit$ yet, the search is allowed to continue until a first sequence is found (or up to $RemainingTime()$).

In the end, the solution is found using the knowledge of the sequences on each machine $M_i \in M$.

Algorithm 1. Solving the decomposed model.

---

   function SOLVEDECOMPOSED
      for each $M_i \in M$ **do**
         $TimeLimit \leftarrow RemainingTime()/(m - i + 2)$
         $Seq_i \leftarrow \text{SEQ}(i, TimeLimit)$
      end for
      Return SOLVE(Seq, $RemainingTime()$)
   end function

Clearly, this decomposition may lead to a schedule arbitrarily far from the optimum, as shown in Sect. 3.1. Hence, we apply the improving phase, that explores other sequences on the machines.

6.2   Improving Phase

Once we have some solution to the problem, the idea of the heuristic is to improve it applying the techniques known as local search [7] and large neighborhood search [12].

It is clear that in order to improve the solution, something needs to be changed on the *critical path*, which is such a sequence of setups and tasks on machines that the completion time of the last task is equal to the makespan and that none of these tasks and setups can be shifted to the left without violating resource constraints (see an example in Fig. 7). Hence, we find the critical path first.

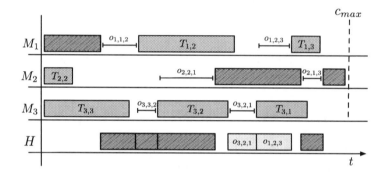

Fig. 7. Illustration of a critical path depicted by dashed rectangles [17].

The most promising place to be changed on the critical path could be the longest setup. Hence, we find the longest setup on the critical path, then we prohibit the two consecutive tasks corresponding to the setup from being processed in a row again and re-optimize the sequence on the machine in question. Two tasks are precluded from following one another by setting the corresponding setup time to infinite value. Also, we add extra constraint restricting the makespan to be less than the incumbent best objective value found. The makespan on one machine being equal to or greater than the incumbent best objective value found cannot lead to a better solution.

After a new sequence is found, the solution to the whole problem is again re-optimized subject to the new sequence. The algorithm continues this way until the sequence re-optimization returns infeasible, which happens due to the extra constraint restricting the makespan. It means that the solution quality deteriorated too much and it is unlikely to find a better solution locally at this state. Thus, the algorithm reverts to the initial solution obtained from the decomposed model, restores the original setup times matrices, and tries to prohibit another setup time on the critical path. For this purpose, the list of *nogoods* to be tried is computed once from the first critical path, which is just a list of setups on the critical path sorted in non-increasing order of their lengths. The whole iterative process is repeated until the total time limit is exceeded or all the nogoods are tried.

The entire heuristic algorithm is hereafter referred to as LOFAS (Local Optimization for Avoided Setup). The pseudocode is given in Algorithm 2.

Algorithm 2. Local Optimization for Avoided Setup [17].

---

function LOFAS
    $S^{init} \leftarrow$ SOLVEDECOMPOSED
    $S^{best} \leftarrow S^{init}$
    $P_{crit} \leftarrow$ critical path in S^{init}
    $nogoods \leftarrow \{h_k \in H \cap P_{crit}\}$
    sort $nogoods$ in non-increasing order of lengths
    for each $h_k \in nogoods$ **do**
        $h_{k'} \leftarrow h_k$
        while true **do**
            $(M_i, T_{i,j}, T_{i,j'}) \leftarrow st(h_{k'})$
            $o_{i,j,j'} \leftarrow \infty$
            impose constraint: $\max_{T_{i,j} \in T^{(i)}} C_{i,j} < ObjVal(S^{best})$
            $Seq_i \leftarrow$ SEQ(i, $RemainingTime()/2$)
            if Seq_i is $infeasible$ **then**
                Revert to S^{init}
                Restore original $O^{(i)}, \forall M_i \in M$
                break
            end if
            $S^{new} \leftarrow$ SOLVE(Seq, $RemainingTime()$)
            if $ObjVal(S^{best}) > ObjVal(S^{new})$ **then**
                $S^{best} \leftarrow S^{new}$
            end if
            if $RemainingTime() \leq 0$ **then**
                return S^{best}
            end if
            $P_{crit} \leftarrow$ critical path in S^{new}
            $h_{k'} \leftarrow$ longest setup $\in \{h_k \in H \cap P_{crit}\}$
        end while
    end for
    return S^{best}
end function

Preliminary experiments confirmed the well-known facts that ILP using lazy approach is very efficient for searching an optimal sequence on one resource, and CP is more efficient for minimizing makespan when the lengths of interval variables and the precedences are fixed. Nevertheless, for instances with many tasks, the solution involving ILP might be computationally infeasible. Hence, we also propose to find a suboptimal sequence for each machine by a heuristic method. Thus, in what follows, we distinguish the following two variants of the algorithm

1. **Exact Subproblem.** The sequence is found by ILP with lazy subtour elimination, as described in [17]. In the experiments below, we denote this variant as *LOFAS*.

2. **Heuristic Subproblem.** The suboptimal sequence is found by *Guided Local Search* algorithm implemented in Google's *OR-Tools* [1]. Note that this algorithm is a metaheuristic, hence, it consumes all the time limit assigned even if the objective value is not improving for several iterations (i.e., cannot prove optimality). In the experiments, we denote this variant as *LOFAS/w heur.*

7   Experimental Results

For the implementation of the constraint programming approaches, we used the IBM CP Optimizer version 12.9 [9]. The only parameter that we adjusted is `Workers`, which is the number of threads the solver can use and which we set to 1. In Google's *OR-Tools*, we only set `LocalSearchMetaheuristic` to `GuidedLocalSearch`.

For the integer programming approach, we used Gurobi solver version 8.1 [6]. The parameters that we adjust are `Threads`, which we set to 1, and `MIPFocus`, which we set to 1 in order to make the solver focus more on finding solutions of better quality rather than proving optimality. We note that parameters tuning with Gurobi Tuning Tool did not produce better values over the baseline ones.

The experiments were run on a Dell PC with an Intel® Core™ i7-4610M processor running at 3.00 GHz with 16 GB of RAM. We used a time limit of 60 s per problem instance.

7.1   Problem Instances

We evaluated the approaches on randomly generated instances of various sizes with the number of machines m ranging from 1 to 50 and the number of tasks on each machine $n_i = n$, $\forall M_i \in M$, ranging from 2 to 50. Thus, we generated $50 \times 49 = 2450$ instances in total. Processing times of all the tasks and setup times are chosen uniformly at random from the interval $[1, 50]$. Instances are publicly available at https://github.com/CTU-IIG/NonOverlappingSetupsScheduling.

7.2   Scalability with Respect to Machines and Tasks

Figure 8a shows the dependence of the best objective value found by CP models within the 60s time limit on the number of machines, averaged over the various number of tasks. Analogically, Fig. 8b shows the dependence of the best objective value on the number of tasks, averaged over the varying number of machines.

The results show that the performances of CP models are almost equal (the graphs almost amalgamate). However, it can be seen that the curve of CP5 is not complete. It is because CP5 fails to find any solution for 88 of the largest instances, despite having the lowest number of variables. We note that CP3 is the best on average but the advantage is negligible. On the other hand, CP4 has better worst-case performance. We will no longer distinguish between the CP models and we will use the minimum of all five CP models that will be referred to as *CPmin*.

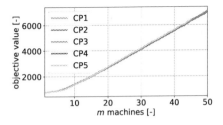
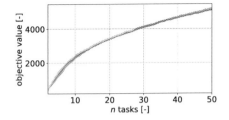

(a) Mean objective value for different number of machines m.   (b) Mean objective value for different number of tasks n.

Fig. 8. Comparison of CP models.

The comparison of CP3ws to ILPws is shown in Fig. 9. Recall that both the approaches get a warm start in a certain sense. The results confirm lower performance of the ILP approach. When the ILP approach model did not get the initial solution as a warm start, it was not able to find any solution even for very small instances (i.e., 2 machines and 8 tasks). In fact, the objective value found by the ILPws is often the objective value of the greedy initial solution given as the warm start (i.e., Sect. 4.1).

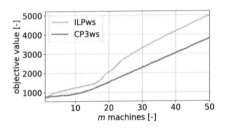
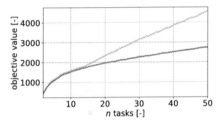

(a) Mean objective value for different number of machines m.   (b) Mean objective value for different number of tasks n.

Fig. 9. Comparison of exact models with warm starts.

Further, we compare the best objective value found by the heuristic algorithm LOFAS and LOFAS /w heur. from Sect. 6 against CPmin and ILPws. The results are shown in Fig. 10. Note that we omit the results of CP3ws (CP3 model with warm starts) in Fig. 10 as the results were almost the same as those of LOFAS and the curves amalgamated.

To obtain better insight into the performance of the proposed methods, we compared the resulting distributions of achieved objectives from each method. We took results of each method for all instances and ordered them in a non-decreasing way with respect to achieved objective value and plotted them. The results are displayed in Fig. 11. It can be seen that the proposed heuristics are able to find the same or better solutions in nearly all cases. Furthermore, we note that LOFAS /w heur. outperformed both CPmin and ILPws as well. For the ILPws, one can notice a spike at around 65% of instances. This is caused by the fact that for some instances, the ILP solver was not able to improve upon

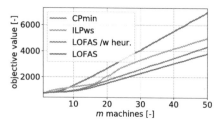 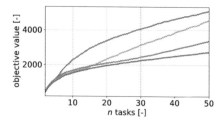

(a) Mean objective value for different number of machines m.   (b) Mean objective value for different number of tasks n.

Fig. 10. Comparison of exact models and the heuristic algorithms.

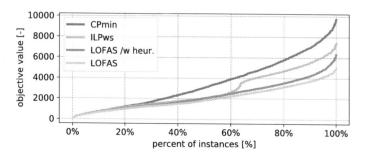

Fig. 11. Objective distributions of different methods.

the initial warm start solution in the given time limit and these instances thus contribute to the distribution with higher objective values.

A comparison of LOFAS to CP3ws on larger instances [17] showed the superiority of LOFAS. However, LOFAS still did not find any solution to the biggest instances (as reported in [17]) because the time limit was exceeded during the decomposition phase, i.e., during seeking an optimal sequence for a machine. This was the motivation for developing LOFAS */w heur.*

7.3   Comparison of Exact and Heuristic Subproblem in LOFAS

In this section, we compare solution quality produced by LOFAS heuristics with different methods for solving the machine subproblem, i.e., the method SEQ(i, *TimeLimit*). The experiments were designed to asses if and how much different objective values are achieved when using the heuristic solution of the subproblem and if the heuristic variant scales better.

We have generated instances with $m \in \{5, 10, 15, 20\}$ machines and the number of tasks on each machine n ranging from 50 up to 1000. For each combination of m and n, we have generated 10 instances. The results are reported in Tab. 1. The column *objective* denotes the mean objective value if all instances were solved in the given time limit. Otherwise, we report the number of instances that were solved within the time limit. The results confirm the hypothesis that LOFAS gives better solutions regarding the objective, whereas LOFAS */w heur.* is able to find some solutions for larger instances when LOFAS does not manage to find any solution. More precisely, LOFAS scales only up to 300 tasks on

15 machines or 350 tasks on 5 machines. A heuristic solution of the subproblem in LOFAS /w heur. allows obtaining a solution for instances of up to 1000 tasks on 5 machines. However, with the increasing number of machines, the CP solver struggles to produce any feasible solution to the whole problem, given the sequences on machines. Therefore, LOFAS /w heur. did not find a solution for any instance with 20 machines, starting from 400 tasks.

To provide further details into the behavior of the algorithm, we report two other statistics. *Dead ends* shows how many times the algorithm hit an infeasible subproblem (due to the constraint on the objective) and restarted to the original solution, and *improv.* reports the number of iterations, where one iteration means avoiding the largest setup on the critical path and solving the subproblem. It can be clearly seen that these two numbers are significantly lower for LOFAS /w heur. because it almost always wastes all the time allocated to a subproblem, whereas LOFAS is able to save some time on smaller instances, which is efficiently used for exploring more potential improvements. Note that for the instances that were not solved, these numbers are always zero.

Table 1. Comparison of exact and heuristic subproblem solvers in LOFAS heuristics.

m	n	LOFAS with exact subproblem			LOFAS with heuristic subproblem		
		objective [-]	dead ends [-]	improv. [-]	objective [-]	dead ends [-]	improv. [-]
5	50	1504.1 ($\pm$57.8)	64.6 ($\pm$2.4)	134.3 ($\pm$51.4)	1517.9 ($\pm$55.2)	0.0 ($\pm$0.0)	4.9 ($\pm$1.8)
10	50	1514.9 ($\pm$49.4)	91.4 ($\pm$43.4)	349.1 ($\pm$145.9)	1545.9 ($\pm$32.2)	0.0 ($\pm$0.0)	0.8 ($\pm$1.2)
15	50	1623.3 ($\pm$48.3)	0.0 ($\pm$0.0)	15.0 ($\pm$31.6)	1871.9 ($\pm$24.3)	0.0 ($\pm$0.0)	0.0 ($\pm$0.0)
20	50	2012.1 ($\pm$29.4)	0.0 ($\pm$0.0)	3.9 ($\pm$6.0)	2485.8 ($\pm$45.2)	0.0 ($\pm$0.0)	1.6 ($\pm$0.7)
5	100	2874.9 ($\pm$88.7)	72.2 ($\pm$46.2)	247.3 ($\pm$127.6)	2939.6 ($\pm$83.1)	0.0 ($\pm$0.0)	4.8 ($\pm$2.5)
10	100	2982.7 ($\pm$92.9)	67.9 ($\pm$52.0)	189.3 ($\pm$108.1)	3063.3 ($\pm$70.1)	0.0 ($\pm$0.0)	2.3 ($\pm$1.8)
15	100	2879.6 ($\pm$59.5)	17.9 ($\pm$17.1)	77.0 ($\pm$28.4)	3279.2 ($\pm$34.7)	0.0 ($\pm$0.0)	0.0 ($\pm$0.0)
20	100	2985.8 ($\pm$32.0)	0.0 ($\pm$0.0)	15.6 ($\pm$19.7)	4035.0 ($\pm$50.9)	0.0 ($\pm$0.0)	0.0 ($\pm$0.0)
5	150	4162.4 ($\pm$156.2)	66.5 ($\pm$36.1)	125.6 ($\pm$62.5)	4258.9 ($\pm$150.6)	0.0 ($\pm$0.0)	2.4 ($\pm$3.1)
10	150	4309.0 ($\pm$128.0)	25.5 ($\pm$16.8)	79.3 ($\pm$40.0)	4415.2 ($\pm$120.9)	0.0 ($\pm$0.0)	0.7 ($\pm$1.3)
15	150	4315.1 ($\pm$95.0)	6.4 ($\pm$7.2)	43.6 ($\pm$13.8)	4673.6 ($\pm$72.0)	0.0 ($\pm$0.0)	0.0 ($\pm$0.0)
20	150	4348.3 ($\pm$44.1)	5.6 ($\pm$7.3)	22.3 ($\pm$7.3)	5456.4 ($\pm$60.1)	0.0 ($\pm$0.0)	0.0 ($\pm$0.0)
5	200	5612.8 ($\pm$130.1)	39.6 ($\pm$22.8)	67.8 ($\pm$26.3)	5707.2 ($\pm$126.7)	0.0 ($\pm$0.0)	5.1 ($\pm$1.9)
10	200	5629.6 ($\pm$64.2)	5.6 ($\pm$10.2)	30.9 ($\pm$18.2)	5734.3 ($\pm$56.6)	0.0 ($\pm$0.0)	1.4 ($\pm$1.5)
15	200	5677.4 ($\pm$112.7)	7.6 ($\pm$5.4)	25.4 ($\pm$6.9)	6032.4 ($\pm$64.7)	0.0 ($\pm$0.0)	0.0 ($\pm$0.0)
20	200	5674.0 ($\pm$92.6)	3.4 ($\pm$3.6)	10.8 ($\pm$3.1)	7023.8 ($\pm$61.8)	0.0 ($\pm$0.0)	0.0 ($\pm$0.0)
5	250	6803.2 ($\pm$115.6)	27.6 ($\pm$12.4)	40.7 ($\pm$9.4)	6903.8 ($\pm$115.6)	0.0 ($\pm$0.0)	4.8 ($\pm$0.4)
10	250	6958.9 ($\pm$131.7)	5.8 ($\pm$7.2)	19.0 ($\pm$5.5)	7100.1 ($\pm$125.5)	0.0 ($\pm$0.0)	0.6 ($\pm$1.3)
15	250	7011.9 ($\pm$177.1)	1.3 ($\pm$1.5)	7.2 ($\pm$3.9)	7417.1 ($\pm$45.5)	0.0 ($\pm$0.0)	0.2 ($\pm$0.4)
20	250	6999.8 ($\pm$58.8)	0.6 ($\pm$0.8)	2.1 ($\pm$1.6)	8339.4 ($\pm$107.6)	0.0 ($\pm$0.0)	0.0 ($\pm$0.0)
5	300	8129.7 ($\pm$99.2)	15.2 ($\pm$9.1)	20.9 ($\pm$11.6)	8246.7 ($\pm$91.4)	0.0 ($\pm$0.0)	2.3 ($\pm$2.5)
10	300	8237.6 ($\pm$92.3)	4.7 ($\pm$4.1)	8.6 ($\pm$4.2)	8384.7 ($\pm$54.9)	0.0 ($\pm$0.0)	0.6 ($\pm$1.0)
15	300	8361.0 ($\pm$61.4)	2.4 ($\pm$2.4)	5.1 ($\pm$2.8)	8707.8 ($\pm$61.1)	0.0 ($\pm$0.0)	0.0 ($\pm$0.0)
20	300	5/10	0.3 ($\pm$0.5)	0.4 ($\pm$0.5)	9883.1 ($\pm$131.9)	0.0 ($\pm$0.0)	0.0 ($\pm$0.0)
5	350	9719.8 ($\pm$177.5)	14.6 ($\pm$5.5)	16.3 ($\pm$4.3)	9822.0 ($\pm$177.5)	0.0 ($\pm$0.0)	4.6 ($\pm$0.5)
10	350	6/10	4.2 ($\pm$4.1)	4.6 ($\pm$4.3)	9785.4 ($\pm$51.6)	0.0 ($\pm$0.0)	1.1 ($\pm$1.2)
15	350	2/10	0.1 ($\pm$0.3)	0.1 ($\pm$0.3)	10417.9 ($\pm$96.9)	0.0 ($\pm$0.0)	0.0 ($\pm$0.0)
20	350	0/10	0.0 ($\pm$0.0)	0.0 ($\pm$0.0)	2/10	0.0 ($\pm$0.0)	0.0 ($\pm$0.0)

(continued)

Table 1. (*continued*)

m	n	LOFAS with exact subproblem			LOFAS with heuristic subproblem		
		objective [–]	dead ends [–]	improv. [–]	objective [–]	dead ends [–]	improv. [–]
5	400	8/10	5.0 (±3.4)	6.0 (±4.3)	11234.6 (±96.3)	0.0 (±0.0)	3.4 (±1.3)
10	400	0/10	0.0 (±0.0)	0.0 (±0.0)	11113.3 (±113.8)	0.0 (±0.0)	0.4 (±0.5)
15	400	0/10	0.0 (±0.0)	0.0 (±0.0)	11770.8 (±85.4)	0.0 (±0.0)	0.0 (±0.0)
20	400	0/10	0.0 (±0.0)	0.0 (±0.0)	0/10	0.0 (±0.0)	0.0 (±0.0)
5	450	8/10	3.2 (±4.7)	3.2 (±4.7)	12276.4 (±157.3)	0.0 (±0.0)	3.2 (±1.7)
10	450	0/10	0.0 (±0.0)	0.0 (±0.0)	12498.1 (±131.5)	0.0 (±0.0)	1.0 (±0.7)
15	450	0/10	0.0 (±0.0)	0.0 (±0.0)	13095.3 (±77.9)	0.0 (±0.0)	0.0 (±0.0)
20	450	0/10	0.0 (±0.0)	0.0 (±0.0)	0/10	0.0 (±0.0)	0.0 (±0.0)
5	500	3/10	0.8 (±1.3)	1.0 (±1.7)	13910.3 (±226.5)	0.0 (±0.0)	3.3 (±1.3)
10	500	0/10	0.0 (±0.0)	0.0 (±0.0)	13888.2 (±188.1)	0.0 (±0.0)	0.7 (±0.8)
15	500	0/10	0.0 (±0.0)	0.0 (±0.0)	7/10	0.0 (±0.0)	0.0 (±0.0)
20	500	0/10	0.0 (±0.0)	0.0 (±0.0)	0/10	0.0 (±0.0)	0.0 (±0.0)
5	550	0/10	0.0 (±0.0)	0.0 (±0.0)	15077.0 (±150.9)	0.2 (±0.4)	2.9 (±0.7)
10	550	0/10	0.0 (±0.0)	0.0 (±0.0)	15196.8 (±197.6)	0.0 (±0.0)	0.2 (±0.4)
15	550	0/10	0.0 (±0.0)	0.0 (±0.0)	1/10	0.0 (±0.0)	0.0 (±0.0)
20	550	0/10	0.0 (±0.0)	0.0 (±0.0)	0/10	0.0 (±0.0)	0.0 (±0.0)
5	600	0/10	0.0 (±0.0)	0.0 (±0.0)	16528.5 (±260.3)	0.2 (±0.4)	3.0 (±1.2)
10	600	0/10	0.0 (±0.0)	0.0 (±0.0)	16706.7 (±327.2)	0.0 (±0.0)	0.2 (±0.4)
15	600	0/10	0.0 (±0.0)	0.0 (±0.0)	0/10	0.0 (±0.0)	0.0 (±0.0)
20	600	0/10	0.0 (±0.0)	0.0 (±0.0)	0/10	0.0 (±0.0)	0.0 (±0.0)
5	650	0/10	0.0 (±0.0)	0.0 (±0.0)	17676.3 (±172.8)	0.2 (±0.4)	1.9 (±1.7)
10	650	0/10	0.0 (±0.0)	0.0 (±0.0)	18235.4 (±173.0)	0.0 (±0.0)	0.2 (±0.4)
15	650	0/10	0.0 (±0.0)	0.0 (±0.0)	0/10	0.0 (±0.0)	0.0 (±0.0)
20	650	0/10	0.0 (±0.0)	0.0 (±0.0)	0/10	0.0 (±0.0)	0.0 (±0.0)
5	700	0/10	0.0 (±0.0)	0.0 (±0.0)	19114.6 (±317.8)	0.4 (±0.8)	3.2 (±0.4)
10	700	0/10	0.0 (±0.0)	0.0 (±0.0)	19532.0 (±198.8)	0.0 (±0.0)	0.0 (±0.0)
15	700	0/10	0.0 (±0.0)	0.0 (±0.0)	0/10	0.0 (±0.0)	0.0 (±0.0)
20	700	0/10	0.0 (±0.0)	0.0 (±0.0)	0/10	0.0 (±0.0)	0.0 (±0.0)
5	750	0/10	0.0 (±0.0)	0.0 (±0.0)	20477.8 (±296.6)	0.1 (±0.3)	2.9 (±0.6)
10	750	0/10	0.0 (±0.0)	0.0 (±0.0)	20774.5 (±123.7)	0.0 (±0.0)	0.2 (±0.4)
15	750	0/10	0.0 (±0.0)	0.0 (±0.0)	0/10	0.0 (±0.0)	0.0 (±0.0)
20	750	0/10	0.0 (±0.0)	0.0 (±0.0)	0/10	0.0 (±0.0)	0.0 (±0.0)
5	800	0/10	0.0 (±0.0)	0.0 (±0.0)	21762.7 (±179.1)	1.2 (±1.2)	2.8 (±2.1)
10	800	0/10	0.0 (±0.0)	0.0 (±0.0)	8/10	0.1 (±0.3)	0.5 (±0.7)
15	800	0/10	0.0 (±0.0)	0.0 (±0.0)	0/10	0.0 (±0.0)	0.0 (±0.0)
20	800	0/10	0.0 (±0.0)	0.0 (±0.0)	0/10	0.0 (±0.0)	0.0 (±0.0)
5	850	0/10	0.0 (±0.0)	0.0 (±0.0)	23144.0 (±210.5)	1.5 (±1.8)	2.8 (±2.6)
10	850	0/10	0.0 (±0.0)	0.0 (±0.0)	9/10	0.0 (±0.0)	0.0 (±0.0)
15	850	0/10	0.0 (±0.0)	0.0 (±0.0)	0/10	0.0 (±0.0)	0.0 (±0.0)
20	850	0/10	0.0 (±0.0)	0.0 (±0.0)	0/10	0.0 (±0.0)	0.0 (±0.0)
5	900	0/10	0.0 (±0.0)	0.0 (±0.0)	24555.4 (±316.8)	1.1 (±1.5)	2.7 (±1.8)
10	900	0/10	0.0 (±0.0)	0.0 (±0.0)	1/10	0.0 (±0.0)	0.0 (±0.0)
15	900	0/10	0.0 (±0.0)	0.0 (±0.0)	0/10	0.0 (±0.0)	0.0 (±0.0)
20	900	0/10	0.0 (±0.0)	0.0 (±0.0)	0/10	0.0 (±0.0)	0.0 (±0.0)
5	950	0/10	0.0 (±0.0)	0.0 (±0.0)	25701.5 (±188.7)	0.5 (±0.7)	2.2 (±1.1)
10	950	0/10	0.0 (±0.0)	0.0 (±0.0)	3/10	0.0 (±0.0)	0.0 (±0.0)
15	950	0/10	0.0 (±0.0)	0.0 (±0.0)	0/10	0.0 (±0.0)	0.0 (±0.0)
20	950	0/10	0.0 (±0.0)	0.0 (±0.0)	0/10	0.0 (±0.0)	0.0 (±0.0)
5	1000	0/10	0.0 (±0.0)	0.0 (±0.0)	27054.0 (±275.9)	1.7 (±1.6)	3.3 (±1.9)
10	1000	0/10	0.0 (±0.0)	0.0 (±0.0)	1/10	0.0 (±0.0)	0.0 (±0.0)
15	1000	0/10	0.0 (±0.0)	0.0 (±0.0)	0/10	0.0 (±0.0)	0.0 (±0.0)
20	1000	0/10	0.0 (±0.0)	0.0 (±0.0)	0/10	0.0 (±0.0)	0.0 (±0.0)

7.4 Discussion

We have seen that performances of CP models are almost equal with CP3 being the best but its advantage is almost negligible. Further, the experiments have shown that ILP without a warm start cannot find a feasible solution for instances with $n \geq 8$ tasks reliably, whereas with warm starts it was significantly better than the best CP model without a warm start. The quality of the solutions from CP with warm starts is much better than ILP with warm starts (even though the warm start for CP is not a complete solution), as can be seen in Fig. 9. As expected, the heuristic algorithm LOFAS produced the best solutions among all compared methods, although only slightly better than CP3 model with warm starts. Smaller instances evidenced that LOFAS achieves objective values quite close to optimal ones. The advantage of LOFAS /w heur. can be seen in its scalability capabilities as it can solve instances with up to 1000 tasks on 5 machines. On the 50×49 instance set from Sect. 7.1, LOFAS /w heur. rendered solutions of objective value worse on average by 10.5 % than LOFAS.

8 Conclusions

This paper tackled the problem of scheduling sequence-dependent non-overlapping setups on dedicated machines. An ILP model, five CP models, and a heuristic approach were proposed. The results showed that all exact methods are finding solutions far from optima within the given time limit, whereas the proposed heuristic algorithm finds high-quality solutions in very short computation time.

The main contributions of this paper with respect to [17] are new CP models using the cumulative function and a new complexity result for the restricted version of the problem. Furthermore, we have proposed an enhancement to the LOFAS algorithm proposing a heuristic for the subproblem, which allows solving larger instances, and extensive experimental evaluation that showed the effect of the subproblem solution method on the scalability and solution quality.

For future work, we will consider a more complex problem, which will avoid the limitation that the tasks are already assigned to machines. Also, instead of non-overlapping setups for one machine setter, we will consider more machine setters that will be treated as a resource with limited capacity.

Acknowledgements. We would like to thank Philippe Laborie for his help with the design of CP4 model.

References

1. Google's or-tools. https://developers.google.com/optimization/. Accessed 22 May 2019
2. Allahverdi, A., Ng, C., Cheng, T.E., Kovalyov, M.Y.: A survey of scheduling problems with setup times or costs. Eur. J. Oper. Res. **187**(3), 985–1032 (2008)

3. Applegate, D., Cook, W.: A computational study of the job-shop scheduling problem. ORSA J. Comput. **3**(2), 149–156 (1991)
4. Balas, E.: Project scheduling with resource constraints. Technical report, Carnegie-Mellon University, Pittsburgh, Pa, Management Sciences Research Group (1968)
5. Chen, D., Luh, P.B., Thakur, L.S., Moreno Jr., J.: Optimization-based manufacturing scheduling with multiple resources, setup requirements, and transfer lots. IIE Trans. **35**(10), 973–985 (2003)
6. Gurobi: Constraints. http://www.gurobi.com/documentation/8.1/refman/constraints.html (2019). Accessed 12 June 2019
7. Hentenryck, P.V., Michel, L.: Constraint-Based Local Search. The MIT Press, Cambridge (2009)
8. Laborie, P., Rogerie, J., Shaw, P., Vilim, P.: Reasoning with conditional time-intervals. Part II: an algebraical model for resources. In: FLAIRS Conference, pp. 201–206 (2009)
9. Laborie, P., Rogerie, J., Shaw, P., Vilim, P.: IBM ILOG CP optimizer for scheduling. Constraints **23**(2), 210–250 (2018)
10. Lasserre, J.B., Queyranne, M.: Generic scheduling polyhedra and a new mixed-integer formulation for single-machine scheduling. In: Proceedings of the 2nd IPCO (Integer Programming and Combinatorial Optimization) Conference, pp. 136–149 (1992)
11. Lee, Y.H., Pinedo, M.: Scheduling jobs on parallel machines with sequence-dependent setup times. Eur. J. Oper. Res. **100**(3), 464–474 (1997)
12. Pisinger, D., Ropke, S.: Large neighborhood search. In: Gendreau, M., Potvin, J.Y. (eds.) Handbook of Metaheuristics. International Series in Operations Research & Management Science, vol. 146, pp. 399–419. Springer, Boston (2010). https://doi.org/10.1007/978-1-4419-1665-5_13
13. Ruiz, R., Andres-Romano, C.: Scheduling unrelated parallel machines with resource-assignable sequence-dependent setup times. Int. J. Adv. Manuf. Technol. **57**(5–8), 777–794 (2011)
14. Tempelmeier, H., Buschkuhl, L.: Dynamic multi-machine lotsizing and sequencing with simultaneous scheduling of a common setup resource. Int. J. Prod. Econ. **113**(1), 401–412 (2008)
15. Vallada, E., Ruiz, R.: A genetic algorithm for the unrelated parallel machine scheduling problem with sequence dependent setup times. Eur. J. Oper. Res. **211**(3), 612–622 (2011)
16. Vilim, P., Bartak, R., Cepek, O.: Extension of O(n log n) filtering algorithms for the unary resource constraint to optional activities. Constraints **10**(4), 403–425 (2005). https://doi.org/10.1007/s10601-005-2814-0
17. Vlk., M., Novak., A., Hanzalek., Z.: Makespan minimization with sequence-dependent non-overlapping setups. In: Proceedings of the 8th International Conference on Operations Research and Enterprise Systems - Volume 1: ICORES, pp. 91–101. INSTICC, SciTePress (2019). https://doi.org/10.5220/0007362700910101
18. Wikum, E.D., Llewellyn, D.C., Nemhauser, G.L.: One-machine generalized precedence constrained scheduling problems. Oper. Res. Lett. **16**(2), 87–99 (1994). http://www.sciencedirect.com/science/article/pii/0167637794900647
19. Zhao, X., Luh, P.B., Wang, J.: Surrogate gradient algorithm for lagrangian relaxation. J. Optim. Theory Appl. **100**(3), 699–712 (1999)

Stochastic Dynamic Pricing with Waiting and Forward-Looking Consumers

Rainer Schlosser[(⊠)]

Hasso Plattner Institute, University of Potsdam, Potsdam, Germany
`rainer.schlosser@hpi.de`

Abstract. Many customers act strategically by checking offer prices multiple times and by anticipating future prices in order to optimize their surplus. In this context, customers base the timing of their purchase decisions on historical reference prices and their individual reservation prices. As customers' reservation prices are not observable it is challenging for sellers to derive optimized pricing strategies. We present a stochastic dynamic finite horizon framework to compute price adjustments that account for strategic customers, that recur and anticipate future prices using observed reference prices. We show how iteratively derive reference prices that are consistent with the observed offer prices of a seller's optimized feedback pricing strategy. We study how expected profits and the evolution of sales are affected by different strategic behaviors. We find that, on average, a recurring customer behavior leads to higher average prices, delayed sales, and increased profits. For the seller, the presence of forward-looking customers has opposing but less intense effects. For the customers, we find that recurring behaviors are not beneficial but can be overcompensated by forward-looking behaviors.

Keywords: Dynamic pricing · Strategic customers · Price anticipations · Waiting customers · Reference prices · Social efficiency

1 Introduction

In many markets, it has become easy to adjust and to observe offer prices. Customers repeatedly check prices in order to strategically time their purchase decisions. Furthermore, they also try to anticipate future prices based on historical reference prices.

Practical applications can be found in a variety of contexts, particularly in the case of perishable products or when the sales horizon is limited. Prominent examples are, for instance, the sale of airline tickets, event tickets, accommodation services, fashion goods, and seasonal products.

To compute effective dynamic pricing strategies in the presence of strategic customers is an important problem in revenue management. Practical relevance is high, but the problem appears challenging. The challenge is (i) to account for returning and forward-looking customers, (ii) to include reference prices, and (iii) to derive pricing decisions with acceptable computation times.

© Springer Nature Switzerland AG 2020
G. H. Parlier et al. (Eds.): ICORES 2019, CCIS 1162, pp. 47–69, 2020.
https://doi.org/10.1007/978-3-030-37584-3_3

In this work, we study pricing strategies that take strategic customer behavior into account when updating offer prices. Existing dynamic pricing techniques cannot handle such scenarios efficiently and, hence, force practitioners to limit the scope of their strategies, e.g., by ignoring strategic behavior or by using simple heuristics. Naturally, this limits the potential quality of pricing strategies.

The limitations of existing techniques to account for strategic customers stems from several factors that reflect the challenges behind dynamic pricing: (i) the solution space of strategies is enormous, (ii) a customer's individual willingness-to-pay is not observable, and (iii) customers track the market to strategically *time* their buying decision.

In revenue management models, for simplicity, mostly so-called myopic customers are considered. They simply arrive and decide; they do not return to check prices again and they do not anticipate future prices.

Instead, in our model, we consider the following two sources of strategic customer behavior. First, we allow that customers return with a certain probability in case they refuse to buy, i.e., if their willingness-to-pay (WTP) does not exceed the current offer price. To reflect planning uncertainty over time, we model a customer's future WTP as a random variable. Second, we allow a certain share of customers to anticipate future offer prices (based on the current offer price and predetermined reference prices) as well as their individual future WTP in order to check whether there is an incentive to delay their purchase decision - even if their current WTP exceeds the offer price. This allows customers to optimize their consumer surplus.

We consider a finite horizon model with limited initial inventory (i.e., products cannot be reproduced or reordered). While in the literature demand is often assumed to be of a special highly stylized functional form, we allow for fairly general demand definitions. In our model, demand is characterized by randomized evolutions of individual WTP, which are not observable for the seller. To this end, demand is allowed to generally depend on time, the current offer price, and reference prices.

This paper is an extended version of [9]. The main contribution of [9] is threefold. We (i) present a demand model which is based on individual reservation prices, reference prices, and expected consumer surpluses, (ii) we compute optimized feedback pricing strategies, (iii) we study the impact of different strategic behaviors on the seller's results, and (iv) we propose a Hidden Markov version of our model.

Compared to [9], in this paper, we present extended evaluation studies and make the following additional contributions: First, we show how to iteratively update consumers reference prices in order to make them consistent with the realized offer prices performed by a seller's optimized feedback pricing strategy. Second, we analyze how different consumer behaviors affect the consumer surplus. While we verify that forward-looking customer behavior is beneficial for consumers, we find that recurring customer behavior is harmful for consumers as a returning customer behavior is going to be exploited by the seller by delaying sales at higher prices. We find that a recurring customer behavior reduces consumers' surplus but can be overcompensated by a forward-looking behavior that uses price anticipations.

This paper is organized as follows. In Sect. 2, we discuss related work. In Sect. 3, we describe our model setup and define strategic customer behaviors. In Sect. 4, we present our solution approach and illustrate its results using different numerical examples. In Sect. 5, we evaluate the impact of different customer behaviors. In Sect. 6, we study a HMM version of our model with partially observable states. Conclusions are summarized in the final Sect. 7.

2   Related Work

Selling products is a classical application of revenue management theory, cf., e.g., Talluri and van Ryzin [14], Phillips [7], and Yeoman and McMahon-Beattie [18]. An excellent overview about recent literature in the field of dynamic pricing is given by Chen and Chen [1].

The analysis of the impact of strategic consumer behavior has been studied since Coase [3]. Surveys about strategic customer behavior in revenue management are proposed by Su [13] and Goensch et al. [4]. Further, the recent survey article by Wei and Zhang [15] provides a very detailed overview of publications studying strategic consumers in operations management.

Wei and Zhang [15] distinguish three streams of counteracting strategic customer behavior: (i) pricing, (ii) inventory, and (iii) information. To account for strategic customers via pricing aims to minimize a customer's incentive to wait for future price drops. This includes strategies such as fixed price strategies, pre-announced increasing prices (price commitment strategy), or reimbursements for decreasing prices (price-matching strategy).

The second stream seeks to mitigate strategic waiting behavior by limiting the product availability in order to address a customer's concern of not being able to purchase the product in the future (cf. run-out). Similarly, sellers can also counteract strategic consumer behavior by strategically announcing (or hiding) partial inventory information to highlight the product's scarcity.

As typically observed in practice, in our model, we allow for free price adjustments. On average, however, this leads to comparatively stable price paths. These reference prices can be estimated by strategic customers, cf. Wu et al. [17]. Further models focusing on reference price effects are studied by Popescu and Wu [8], Wu and Wu [16], or Chenavaz and Paraschiv [2].

Finally, the key question is how (i) reference prices, (ii) consumer's price anticipations, and (iii) a firm's pricing strategy affect each other. Further, the mutual dependencies will have to be determined by buyers and sellers based on their partially observable asymmetric (market) data. In addition, the complex interplay of their mutual beliefs is further complicated when multiple seller compete for the same market, cf., e.g., Levin et al. [5] and Liu and Zhang [6].

3   Model Description

We consider the situation in which a firm wants to sell a finite number of goods (e.g., airline tickets, event tickets, accommodation services) over a certain time frame. We assume a monopoly situation. Further, a certain ratio of customers acts strategically, i.e., they (i) repeatedly track prices and wait for acceptable offers and (ii) they are forward-looking, i.e., they compare their current consumer surplus with expected future consumer surpluses.

We assume that the time horizon T is finite. We assume that products cannot be reproduced or reordered. If a sale takes place, shipping costs c have to be paid, $c \geq 0$. A sale of one item at price a, $a \geq 0$, leads to a profit of $a - c$. Discounting is also included in the model. For the length of one period, we use the discount factor δ, $0 \leq \delta \leq 1$.

3.1   Individual Buying Behavior

We consider a discrete time model with T periods. We assume that consumers have individual (random) reservation prices denoted by R_t for periods $(t, t+1)$, $t = 0, 1, ..., T - 1$. The reservation prices particularly account for a customer's planning uncertainty to be able to benefit from the product (e.g., an airline ticket or an event ticket in time T). In this context, the average planning uncertainty typically decreases over time as the time horizon T gets closer.

Further, customers may check current (ticket) prices at multiple points in time. While some customers start to check prices early (cf. economy class) other customers start to check prices shortly before the end of the sales period (cf. business class). Moreover, the evolutions of reservation prices of each individual customer are different. Figure 1 illustrates individual reservation prices R_t and average reservation prices (denoted by $\bar{R}_t$) over time. We observe that, on average, the willingness-to-pay is increasing over time. The individual paths

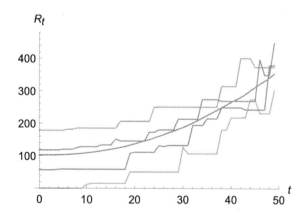

Fig. 1. Examples of individual reservation prices R_t and average reservation prices $\bar{R}_t$ (smooth blue curve) over time, see [9]. (Color figure online)

resemble, e.g., random (sudden) changes of a customer's planning uncertainty. In other use-cases (e.g., the sale of fashion goods) reservation prices may also tend to decrease.

We assume that forward-looking customers compare their current individual consumer surplus with their *expected* future surplus of the next period. Given a current offer price a_t and an individual reservation price R_t at time t, the current consumer surplus is $CS_t := R_t - a_t$. The expected future surplus CS_{t+1} of the next period depends on the (expected) offer price a_{t+1} and the (expected) individual reservation price R_{t+1} of the consumer at time $t + 1$. Finally, a customer purchases a product at time t only if condition (i) is satisfied: The current consumer surplus is positive, i.e., $t = 0, 1, ..., T - 1$,

$$CS_t \geq 0 \tag{1}$$

Further, for some consumer, also a *second* condition (ii) has to be satisfied: The expected future surplus $E(CS_{t+1})$ does not exceed CS_t plus a certain risk premium ε (e.g., mirroring a customer's risk aversion about a product's future availability), so that there is no incentive to wait for a higher consumer surplus), $t = 0, 1, ..., T - 1, \varepsilon \geq 0$,

$$R_t - a_t \geq E(R_{t+1}|R_t) - E(a_{t+1}) - \varepsilon \tag{2}$$

Consumers are assumed to be able to estimate future prices $E(a_{t+1})$ based on average historical reference prices. In this context, in our model, we assume a known predetermined path of reference prices denoted by, $t = 0, 1, ..., T - 1$,

$$a_t^{(ref)} \tag{3}$$

Based on a current offer price a_t and the reference price $a_{t+1}^{(ref)}$, cf. (3), consumers can estimate expected future prices, e.g., by, $t = 0, 1, ..., T - 2$,

$$E(a_{t+1}) \approx (a_{t+1}^{(ref)} + a_t)/2$$

We assume that a certain share of the consumers are forward-looking and consider condition (2). This share is denoted by, $t = 0, 1, ..., T - 1$,

$$\gamma_t \in [0, 1] \tag{4}$$

Finally, given the distribution of the expected evolution of a random customer's reservation price R_t, from (1)–(3), we obtain the average purchase probability of a random customer arriving in period $(t, t + 1)$, $t = 0, 1, ..., T$, $a \geq 0$, $\varepsilon \geq 0$,

$$p_t^{(buy)}(a) := (1 - \gamma_t) \cdot P(R_t > a)$$
$$+ \gamma_t \cdot P \left(\begin{array}{l} R_t > a \quad and \\ R_t - a \geq E(R_{t+1}|R_t) - \frac{a_{t+1}^{(ref)} + a}{2} - \varepsilon \end{array} \right) \tag{5}$$

The purchase probabilities (5) are characterized by (i) the consumers' mixture of reservation prices, (ii) the share of forward-looking customers, cf. (4), and (iii) the reference prices, cf. (3).

Note, we do not assume that reservation prices are observable for the seller. As in practice, we only assume arriving customers and realized sales to be observable for sellers. The probabilities (5) can be estimated by the conversion rate of interested and buying customers for different offer prices at different time t, cf., e.g., Schlosser and Boissier [10, 11].

3.2 Waiting Customers

Typically, customers tend to track the market and observe prices over time. In the literature, however, customers are often assumed to be myopic, i.e., they randomly occur, observe offers, and decide whether to purchase or not. In case of no purchase they do not further track the offer.

In reality, many customers are not myopic. In our model, we consider recurring customers. In case an interested customer does not purchase, we assume that he/she checks the next period's offer with a certain probability denoted by, $t = 0, 1, ..., T - 1$,

$$\eta_t \in [0, 1] \tag{6}$$

Note, this nontrivially affects the arrival process of potential customers. In the following, we distinguish between initially arriving (new) customers and waiting/recurring (old) customers.

Arriving new customers are modelled as follows. We assume arbitrary given probabilities, $t = 0, 1, ..., T - 1$, $j = 0, 1, ...$,

$$p_t^{(new)}(j) \tag{7}$$

that in period $(t, t + 1)$ exactly j new customers arrive; for all $t = 0, 1, ..., T - 1$ we assume $\sum_{j \geq 0} p_t^{(new)}(j) = 1$.

For the time being, we assume that the number of waiting customers can be effectively determined by the selling firm as new arriving customers and old recurring customers can be observed (cf. cookies, etc.). The random number of customers that did not purchase in period $(t - 1, t)$ and recur in the next period $(t, t + 1)$ are denoted by K_t, $t = 0, 1, ..., T - 1$. A list of variables and parameters is given in the Appendix, cf. Table 5.

3.3 Problem Formulation

In our model, we use sales probabilities that depend on (i) the number of arriving new customers j, (ii) the number recurring waiting customers k, and (iii) the offer price a. The individual purchase decisions, cf. (5), are based on individual (expectations of future) reservation prices and predetermined reference prices.

The random inventory level of the seller at time t is denoted by X_t, $t = 0, 1, ..., T$. The end of sale is the random time τ, when all of the seller's items are sold, that is $\tau := \min_{t=0,...,T} \{t : X_t = 0\} \wedge T$. As long as the seller has items left to sell, for each period $(t, t + 1)$, a price a_t has to be chosen. By A we denote the set of admissible prices. For all remaining $t \geq \tau$ the firm cannot sell further items and we let $a_t := 0$.

We call strategies $(a_t)_t$ admissible if they belong to the class of Markovian feedback policies; i.e., pricing decisions $a_t \geq 0$ will depend on (i) time t, (ii) the current inventory level X_t, and (iii) the current number of waiting customers K_t.

Depending on the chosen pricing strategy $(a_t)_t$, the random accumulated profit from time/period t on (discounted on time t) amounts to, $t = 0, 1, ..., T$,

$$G_t := \sum_{s=t}^{T-1} \delta^{s-t} \cdot (a_s(X_s, K_s) - c) \cdot (X_{s+1} - X_s) \tag{8}$$

The objective is to determine a (Markovian) feedback pricing policy that maximizes the expected total discounted profits, $t = 0, 1, ..., T$,

$$E(G_t | X_t = n, K_t = k) \tag{9}$$

conditioned on the current state at time t (cf. inventory level n and waiting consumers k). An optimized policy will balance expected short-term and long-term profits by accounting for the evolution of the inventory level and the number of waiting customers.

4   Computation of Optimal Pricing Strategies with Observable States

In this section, we want to derive optimal feedback pricing strategies that incorporate the strategic customer behavior described in Sect. 2.

4.1   State Transition Probabilities

The state of the system to be controlled over time is described by time t, the current inventory level n, and the current number of waiting customers k. The transition dynamics can be described as follows. Given an inventory level n at time t and a demand for i items during the period $(t, t+1)$, we obtain the new inventory level $X_{t+1} := \max(n - i, 0)$.

Given k waiting customers at time t and j new arriving customers, cf. (7), we have $k + j$ interested customers in period $(t, t+1)$. Assuming i buying customers, $i = 0, ..., k+j$, cf. (1)–(5), we obtain $k + j - i$ customers that did not purchase an item. If m of them plan to recur in $(t+1, t+2)$ the new state, i.e., the number of waiting customers at time $t+1$ is m, $m = 0, ..., k+j-i$.

Assuming k waiting customers and j new customers, the probability that exactly i items can be sold at period $(t, t+1)$ is binomial distributed, $i = 0, ..., k+j$, $t = 0, 1, ..., T-1$, cf. (5),

$$p_t^{(demand)}(i|a,k,j)$$
$$= \binom{k+j}{i} \cdot p_t^{(buy)}(a)^i \cdot \left(1 - p_t^{(buy)}(a)\right)^{k+j-i} \tag{10}$$

Assuming k waiting customers at time t, j new customers, and i customers that want to buy, the probability that m of the remaining $k + j - i$ customers return in period $(t + 1, t + 2)$ is also binomial distributed, $m = 0, ..., k + j - i$, $t = 0, 1, ..., T - 1$, cf. (6),

$$p_t^{(wait)}(m|k,j,i) = \binom{k+j-i}{m} \cdot \eta_t^m \cdot (1 - \eta_t)^{k+j-i-m} \tag{11}$$

4.2   Solution Approach

The problem of finding the best pricing strategy can be solved using dynamic programming techniques. In this context, the so-called value function describes the best expected discounted future profits $E(G_t|n,k)$ for all possible states n and k at time t, cf. (9).

If either all items are sold or the time is up, no future profits can be made, i.e., the natural boundary conditions for the value functions V are given by, $t = 0, 1, ..., T - 1$, $n = 0, 1, ..., N$, $k = 0, 1, ...,$

$$V_t(0, k) = 0 \quad and \quad V_T(n, k) = 0 \tag{12}$$

For all other states the value function is determined by the Bellman equation, $n = 0, 1, ..., N$, $k = 0, 1, ..., M$, $t = 0, 1, ..., T - 1$, cf. (7), (10), (11),

$$V_t(n, k) = \max_{a \in A} \left\{ \sum_{\substack{j=0,1,...,J \\ i=0,1,...,j+k \\ m=0,1,...,j+k-i}} p_t^{(new)}(j) \right. \tag{13}$$
$$\cdot p_t^{(demand)}(i|a,j,k) \cdot p_t^{(wait)}(m|j,k,i)$$
$$\left. \cdot ((a - c) \cdot \min(i, n) + \delta \cdot V_{t+1}(\max(n - i, 0), m)) \right\}$$

Note, to obtain a bounded number of potential events, in (13) we use a maximum number of new customers J. To guarantee a limited state space, we use a maximum number M of waiting customers. Both bounds have to be chosen sufficiently large such that the optimal solution is not confined.

The nonlinear system of equations (13) can be solved recursively. The associated optimal pricing denoted by $a_t^*(n, k)$, $n = 0, 1, ..., N$, $k = 0, 1, ..., M$, $t = 0, 1, ..., T - 1$, is determined by the arg max of (13), i.e.,

$$a_t^*(n,k) = \arg\max_{a \in A} \left\{ \sum_{\substack{j=0,1,\ldots,J \\ i=0,1,\ldots,j+k \\ m=0,1,\ldots,j+k-i}} p_t^{(new)}(j) \right.$$

$$\cdot p_t^{(demand)}(i\,|\,a,j,k) \cdot p_t^{(wait)}(m\,|\,j,k,i)$$

$$\left. \cdot ((a-c) \cdot \min(i,n) + \delta \cdot V_{t+1}(\max(n-i,0),m)) \right\} \tag{14}$$

If $a_t^*(n,k)$ is not unique, we choose the smallest one.

4.3   Numerical Example

To illustrate our solution approach, we consider the following numerical example.

Example 4.1. We let $T = 50$, $N = 20$, $\delta = 1$, $c = 10$, $J = 5$, $M = 8$, $a \in A := \{10, 20, \ldots, 500\}$, $\varepsilon := U(0, 20)$, and $\gamma_t = \eta_t = 0.5$. Further, we use:

(i) Arrival of new customers, cf. (7), $j = 0, 1, \ldots$,

$$p_t^{(new)}(j) = \binom{J}{j} \cdot u_t^j \cdot (1 - u_t)^{J-j}$$

where $u_t = 1 - e^{-|t/T - 0.6|}$, $0 \le t < T$.

(ii) Individual reservation prices, cf. (5), $0 \le t < T$,

$$R_t = \begin{cases} L^{(min)}, & t = 0 \\ if\ U(0,1) < 0.75\ then\ R_{t-1}\ else\ H_t, & t > 0 \end{cases}$$

where $L^{(min)} := U(0, 200)$ (initial reservation price), $L^{(max)} := U(100, 800)$ (upper reservation price), and $H_t := L^{(min)} + D_t \cdot U(0.6, 1.4)$ (random updates) with $D_t := E\left(L^{(max)} - L^{(min)}\right) \cdot (t/T)^2$ (average increase). By $U(\cdot, \cdot)$, we denote Uniform distributions.

(iii) Reference prices, cf. (3), $0 \le t < T$,

$$a_t^{(ref)} := 150 + 200 \cdot (t/T)^2$$

(iv) We use $10\,000$ random realization of R_t, cf. (ii), to determine the average reservation prices $\bar{R}_t := E(R_t)$ and the conditional expectations $E(R_{t+1}|R_t)$, which are the basis to derive $p_t^{(buy)}(a)$, $a \in A$, cf. (5).

Figure 2 depicts purchase probabilities $p_t^{(buy)}(a)$ in the setting of Example 4.1 for different periods t and prices a. Note, the seller cannot infer individual reservation prices from $p_t^{(buy)}$.

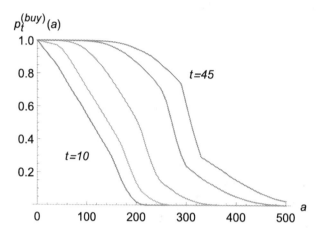

Fig. 2. Purchase probabilities $p_t^{(buy)}(a)$, cf. (5), for different prices $a \leq 500$ and periods $t = 10, 20, 30, 40, 45$; Example 4.1, see [9].

Table 1 illustrates the value function V_t for different inventory levels n and points in time t (for the case that the number of waiting customers is $k = 3$), cf. (13). We observe that expected future profits are (convex) decreasing in time and (concave) increasing in the number of items left to sell.

Table 1. Expected profits $V_t(n, 3)$, Example 4.1, see [9].

$n \backslash t$	0	10	20	30	40	45
1	350	350	350	350	356	356
2	654	654	654	655	655	637
5	1409	1409	1410	1410	1351	1265
10	2413	2413	2414	2383	2037	1475
15	3246	3239	3219	3035	2114	1476
20	3924	3883	3763	3230	2115	1476

Table 2 shows the expected profits for different inventory levels n and different numbers of waiting customers k (for time $t = 45$). We observe that the expected future profits are increasing in the number of waiting consumers k. The impact of waiting customers k is higher the larger the remaining inventory is.

Table 2. Expected profits $V_{45}(n, k)$, Example 4.1, see [9].

$n \backslash k$	0	1	2	3	4	5
1	288	318	339	356	369	380
2	492	552	599	637	667	693
5	744	953	1131	1265	1349	1408
10	762	1000	1238	1475	1711	1942
15	762	1000	1238	1476	1714	1952
20	762	1000	1238	1476	1714	1952

Table 3 illustrates optimal feedback prices a_t^*, cf. (14), for different inventory levels n and points in time t (for the case that the number of waiting customers is $k = 3$), cf. Table 1. We observe that offer prices are decreasing in n and increasing in time t. Further, prices are (slightly) increasing in k. The impact of k is higher if n is small and t is large.

Note, if n and t are small, optimal prices are chosen such that the probability to sell is basically 0, cf. Fig. 2. This way, it is ensured that items are not sold too early. Due to increasing demand (cf. Fig. 1) items can be sold later at higher prices.

Table 3. Optimal prices $a_t^*(n, 3)$, Example 4.1, see [9].

$n \backslash t$	0	10	20	30	40	45
1	200	220	280	370	410	420
2	200	220	280	330	370	380
5	200	220	250	280	280	290
10	200	200	220	230	250	290
15	180	180	190	200	250	290
20	160	160	170	190	250	290

Remark 4.1. (Properties of expected profits)

(i) The expected profits are increasing in the inventory level.
(ii) If there is no discounting then the expected profits are increasing in time-to-go.
(iii) The expected profits are increasing in the number of waiting customers, especially if time-to-go is small and the inventory level is large.

Remark 4.2. (Properties of feedback prices)

(i) The optimal prices are decreasing in the inventory level.

(ii) If demand is strongly increasing in time then the optimal prices are increasing in the time.

(iii) The optimal prices are increasing in the number of waiting customers, especially if time-to-go is small and the inventory level is small.

Figure 3 illustrates evaluated average prices $E(a_t^*)$ (orange curve), which are, on average, increasing over time. Compared to $E(a_t^*)$ the average reservation prices $\bar{R}_t$ (blue curve) are smaller in the beginning and higher at the end of the sales period. Hence, sales are likely to occur late. The green curve depicts the predetermined reference prices $a_t^{(ref)}$, which are overall *consistent* with the evaluated offer prices $E(a_t^*)$.

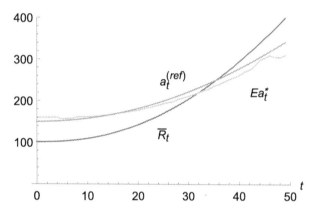

Fig. 3. Average evaluated price paths $E(a_t^*)$ (orange curve), average reservation prices $\bar{R}_t$ (blue curve), and reference prices $a_t^{(ref)}$ (green curve); Example 4.1, see [9]. (Color figure online)

4.4   Iterating Consistent Reference Prices

In this section, we show how to approximate reference prices that are consistent with the optimized prices applied by the firm. The resulting average offer prices set by the firm (cf. $E(a_t^*)$) can be used to define or adjust the expected reference price paths $a_t^{(ref)}$, cf. (3). This way, adaptive reconfigurations and iterated model solutions can be used to approximate converging equilibrium price paths where the evaluated optimized solution ($E(a_t^*)$) coincides with the underlying expectations of customers ($a_t^{(ref)}$).

Following this idea, we compute iterated reference prices denoted by $a_{t,k}^{(ref)}$, where k is the iteration step. We start with a given initial reference price path $a_{t,1}^{(ref)}$, cf. $k=1$. Given the reference prices $a_{t,k}^{(ref)}$, $k \geq 1$, we compute the optimized feedback pricing policy $a_{t,k}^*$ via (13)–(14) and simulate associated average offer prices $E(a_{t,k}^*)$ that occur under a certain consumer behavior (which also includes the current reference prices $a_{t,k}^{(ref)}$). We can assume that consumers update their reference prices by the observed average offer prices $E(a_t^*)$ (social learning). In our model, we describe this update process by, $k \geq 1$, $t = 0, ..., T - 1$,

$$a_{t,k+1}^{(ref)} := \left(a_{t,k}^{(ref)} + E(a_{t,k}^*) \right) / 2 \tag{15}$$

Using updated reference prices a new optimized feedback pricing policy $a_{t,k+1}^*$ can be computed and the simulation and the reference update (15) is applied again. The goal is to iteratively proceed until the reference price and the applied offer prices converge to one consistent path.

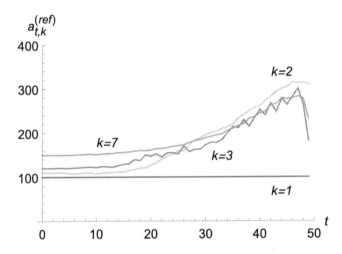

Fig. 4. Convergence of reference prices and offer prices: Iterating reference prices via (15) starting from the initial reference price $a_{t,1}^{(ref)} = 100$, $t = 0, ..., T - 1$; Example 4.1.

To demonstrate the iteration approach and to verify that the equilibrium path does not significantly depend on the initial reference, we used two different initial reference price path $a_{t,1}^{(ref)} = 100$ and $a_{t,1}^{(ref)} = 300$, $t = 0, ..., T - 1$. The results are shown in Figs. 4 and 5 for the setting of Example 4.1.

In both cases, we observe that already after a few iteration steps the reference prices converge to one similar curve, which is also consistent with the reference price curve used in the previous sections, cf. Fig. 3. We can assume that such equilibrium curves are mainly characterized by the average evolution of consumers' reservation prices and the firm's price optimization, cf. (13)–(14).

5 Comparison of Different Customer Behaviors

In this section, we study the impact of different setups of strategic customer behaviors, cf. (4) and (6), on the expected seller's profits and the expected consumer surplus.

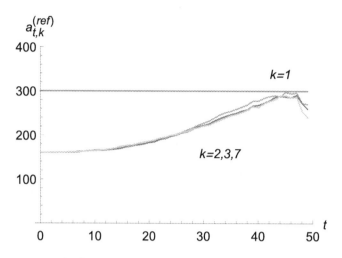

Fig. 5. Convergence of reference prices and offer prices: Iterating reference prices using alternative initial reference prices $a_{t,1}^{(ref)} = 300$, $t = 0, ..., T - 1$; Example 4.1.

5.1   Evaluation of Consumer Surplus

While the evaluation of the expected profits of the seller is straightforward, we evaluate the average consumer surplus as follows. Assume a consumer arriving in period $(t, t + 1)$, $t = 0, ..., T - 1$, decides to buy an item at the current offer price a_t^* then his/her average consumer surplus follows the definition of (5) and amounts to, $t = 0, ..., T - 1$,

$$ECS_t := (1 - \gamma_t) \cdot E\left(R_t - a_t^*(X_t, K_t)|R_t > a_t^*(X_t, K_t)\right)$$

$$+ \gamma_t \cdot E\left(R_t - a_t^*(X_t, K_t) \left| \begin{array}{l} R_t > a_t^*(X_t, K_t) \text{ and} \\ R_t - a_t^*(X_t, K_t) \geq E(R_{t+1}|R_t) \\ -\frac{a_t^{(ref)} + a_t^*(X_t, K_t)}{2} - \varepsilon \end{array} \right. \right) \quad (16)$$

The conditional expectation (15) can, e.g., be evaluated using simulated reservation prices R_t as well as multiple simulated sales processes for a given feedback strategy a^* (which depends on the current number of remaining items X_t and the number of waiting customers K_t).

The expected total consumer surplus (accumulated until time t) is based on (15) and the average number of sales $(E(X_t) - E(X_{t-1}))$ in period $(t - 1, t)$, $t = 1, ..., T$,

$$ECS_t^{(cum)} := \sum_{s=1,...,t} ECS_{s-1} \cdot (E(X_s) - E(X_s - 1)) \quad (17)$$

In the next two subsections, we evaluate results for different strategic customers from a seller's perspective (Sect. 5.2) and the consumers' perspective (Sect. 5.3).

5.2   Evaluation from a Seller's Perspective

In the following, we evaluate the seller's profit as well as the consumer surplus using a numerical example. We consider Example 4.1 for different customer behaviors.

Example 5.1. We assume the setting of Example 4.1. We consider different combinations of (time consistent) parameters $\eta_t = \eta$ (return probability), and $\gamma_t = \gamma$ (share of forward-looking customers) characterizing the customers' strategic behavior.

Table 4 summarizes the seller's expected profits as well as the average total consumer surplus for five different customer behaviors, cf. Example 5.1. As a reference, setting I represents the classical myopic customer behavior without any kind of strategic effects. In setting II, all consumers are forward-looking, cf. (4), but do not recur. In setting III, all consumers recur, cf. (6), but are not forward-looking. In setting IV, all consumers are forward-looking and steadily recur. Setting V corresponds to the mixed setup of Example 4.1.

Table 4. Expected total profits $E(G_0)$ and expected total consumer surpluses $ECS_T^{(cum)}$ for different strategic customer behaviors; Example 5.1.

Setting	η	γ	$E(G_0)$	$ECS_T^{(cum)}$	Sum
I	0	0	3 455	1 148	4 603
II	0	1	3 390	1 197	4 587
III	1	0	6 241	690	6 931
IV	1	1	4 940	1 199	6 139
V	0.5	0.5	3 924	1 036	4 960

Comparing the expected profits in Table 4, we observe that the strategic behavior has a significant impact on both a seller's sales results and the consumer surplus. While for the seller a higher return probability η and a lower anticipation probability γ is beneficial, for the consumers we observe opposed effects. Setting III is the best (worst) one for the seller (consumer) and maximizes the sum of both surpluses (cf. social efficiency). Setting I and II are the worst for the seller. Setting II and IV are the best for the consumer.

Next, we investigate how the different behaviors affect sales processes over time. Figure 6 illustrates the expected price paths associated to the first four settings I–IV, cf. Table 4. While the increasing shape of the four paths is overall similar, the level of prices differs significantly. The highest (lowest) prices correspond to setting III (setting II). The moderate setting V, cf. Example 4.1 and Fig. 2, corresponds to a mixture of setting I and IV, i.e., the blue and the red curve.

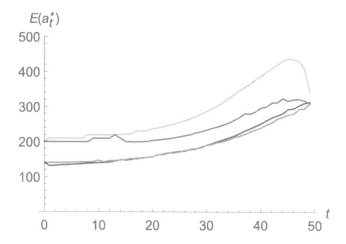

Fig. 6. Expected evolution of optimal prices $E(a_t^*)$ for different strategic behaviors (settings: I blue, II orange, III green, IV red); Example 5.1, see [9]. (Color figure online)

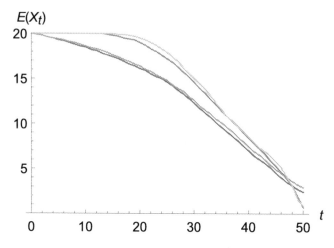

Fig. 7. Expected evolution of the inventory level $E(X_t)$ for different strategic behaviors (settings: I blue, II orange, III green, IV red); Example 5.1, see [9]. (Color figure online)

For setting I–IV, Fig. 7 depicts the associated inventory levels over time. All four curves are of concave shape. Most of the sales are realized at the end of the time horizon, which is due to the increasing demand (i.e., reservation prices). The low return probability of setting I and II (blue and orange curve) leads to sales that occur comparatively early. Instead, in setting III and IV (green and red curve) with recurring customers sales can be realized later and the number of unsold items can be reduced.

Further numerical experiments led to similar results. We summarize our observations for the seller in the following remark.

Remark 5.1. (Impact of strategic behavior for the seller)

(i) Compared to myopic settings a higher share (η) of patient recurring customers leads to higher profits. The (predictable) higher number of interested customers results in higher prices and delayed sales.

(ii) A higher share (γ) of forward-looking customers leads to lower profits. Customers strategical time their purchase in order to increase their consumer surplus. Further, due to suspended decision-making, sales – that would have been realized in myopic settings – may even get lost. Hence, a higher number of anticipating customers forces the seller to lower prices which, in turn, leads to earlier sales.

(iii) Finally, the two effects, i.e., the quantities η and γ, are counteractive. When comparing both effects, we observe that the impact of anticipating customers is overcompensated by those of recurring customers.

5.3   Evaluation from a Consumers' Perspective

In the following, we study how the seller's optimized strategy against different customer behaviors affects the expected consumer surplus. Again, we consider the setting of the numerical Example 5.1.

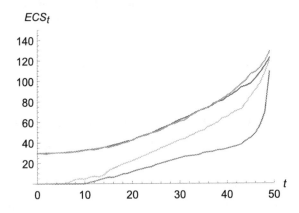

Fig. 8. Expected consumer surplus per sale ECS_t of a sale, cf. (16), over time for different strategic behaviors (settings: I blue, II orange, III green, IV red); Example 5.1. (Color figure online)

Figure 8 shows the evolution of the expected consumer surplus per sale ECS_t of a sale, cf. (16). Figure 9 illustrates the accumulation of the average realized consumer surplus over time.

In setting III (return, no forward-looking), the consumer surplus (per sale) is comparatively low, which is due to the higher prices induced by the corresponding pricing strategy, see Fig. 6. The consumer surplus is comparably high if customers do *not* recur (setting I and II) as the seller can plan expected consumer arrivals with lower confidence.

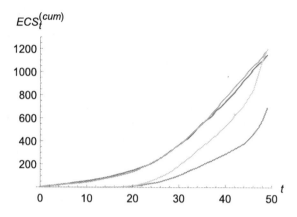

Fig. 9. Expected accumulated consumer surplus $ECS_T^{(cum)}$, cf. (17), over time for different strategic behaviors (settings: I blue, II orange, III green, IV red); Example 5.1. (Color figure online)

Remark 5.2. (Impact of strategic behavior for the consumers)

(i) Compared to myopic settings a higher share (η) of patient recurring customers negatively affects the expected consumer surplus.

(ii) A higher share (γ) of forward-looking customers leads to a higher expected consumer surplus.

(iii) When comparing both effects, we observe that the impact of recurring customers is overcompensated by those of anticipating customers.

6   Pricing Strategies When Waiting Customers Are Not Observable

In real-life applications, the number of returning customers is typically not exactly known. Further, it might not always be possible to distinguish between new and recurring customers. In this section, we show how to adjust the model presented in Sect. 4 by using probability distributions of waiting customers. This makes it possible to still compute effective strategies although less information is available.

6.1   Return Probabilities

While the number of recurring customers is often not observable, the average return probability can be easily estimated. Based on average return probabilities it is possible to derive a probability distribution for the number of recurring customers.

The key idea is to exploit the model with full information, cf. Sect. 4, and to use probabilities for waiting customers (cf. Hidden Markov Model).

The probability that k customers return in period $(t, t+1)$ is denoted by $\pi_t^{(k)} := P_t(K_t = k)$. We can estimate $\pi_t^{(k)}$ based on the observable number of interested customers v, i.e., the sum of old and new customers, cf. $v := k+j$, see Sect. 4.1. Assume in a period $(t-1, t)$, we observed v interested customers and i buyers. Hence, we have, cf. (11), $k = 0, 1, ..., M$, $t = 0, 1, ..., T$, $v = 0, 1, ..., J+M$, $i = 0, 1, ..., v$,

$$\pi_t^{(k)} := P(K_t = k | v, i)$$
$$= \binom{v-i}{k} \cdot \eta_t^k \cdot (1 - \eta_t)^{v-i-k} \tag{18}$$

For all $k > v-i$, we obtain $\pi_t^{(k)} := 0$. Note, as (18) is based on the observable number of interested (v) and buying customers (i), we do not need to be able to distinguish between new and old customers.

6.2   Computation of Prices

Next, we compute optimized strategies. We use given (state) probabilities $\pi_t^{(k)}$, cf. (18), and the value function $V_t(n, k)$, cf. (13), of the scenario with full information, see [12] for a similar HMM approach. We define the following heuristic pricing strategy denoted by $\tilde{a}_t(n)$ for the scenario with unobservable recurring customers, $n = 0, 1, ..., N$, $k = 0, 1, ..., M$, $t = 0, 1, ..., T$,

$$\tilde{a}_t(n) = \arg\max_{a \in A} \left\{ \sum_{k=0,...,M} \pi_t^{(k)} \cdot \sum_{\substack{j=0,1,...,J \\ i=0,1,...,j+k \\ m=0,1,...,j+k-i}} p_t^{(new)}(j) \right.$$
$$\cdot p_t^{(demand)}(i | a, j, k) \cdot p_t^{(wait)}(m | j, k, i) \tag{19}$$
$$\left. \cdot ((a-c) \cdot \min(i, n) + \delta \cdot V_{t+1}(\max(n-i, 0), m)) \right\}$$

Algorithm 6.1. Use (13), (18), and (19) in the following order to compute $\tilde{a}_t(X_t)$, $t = 0, 1, ..., T-1$:

(i) Compute the values $V_t(n, k)$ for all $n = 0, 1, ..., N$, $k = 0, 1, ..., M$, and $t = 0, ..., T$ via (13).

(ii) In $t = 0$ let $\pi_0^{(k)} := 1_{\{k=0\}}$. Compute the price $\tilde{a}_0(N)$ using (19) for the initial inventory $X_0 := N$.

(iii) For all $t = 1, ..., T-1$ observe the number v of interested customers in period $(t-1, t)$ and the number i of buying customers. Given v and i compute $\pi_t^{(k)}$, $k = 0, 1, ..., M$, via (18). Let $X_t = X_{t-1} - i$. Use $\pi_t^{(k)}$ to compute the price $\tilde{a}_t(X_t)$ for the current inventory level X_t, cf. (19).

6.3   Numerical Examples

In the following example, we demonstrate the applicability and the quality of our Hidden Markov approach, cf. Algorithm 6.1.

Example 6.1. We assume the setting of Example 4.1. We assume that recurring customers cannot be observed by the seller. Instead of the pricing strategy (13)–(14), the seller applies the Hidden Markov approach described in Algorithm 6.1. We consider the customer behavior characterized by $\eta_t = 0.5$ and $\gamma_t = 0.5$.

Figure 10 illustrates average evaluated price curves of our heuristic strategy $\tilde{a}_t$ compared to the optimal strategy a_t^*, which, in contrast to the heuristic, takes advantage of being able to observe waiting customers. We observe that both curves are almost identical which indicates that realized prices of both strategies are similar.

Figure 11 shows the corresponding evolutions of accumulated profits up to time t (denoted by $\bar{G}_t$). The curves verify that the performance of the heuristic strategy is close to optimal. For other settings of the customer behavior we obtain similar results.

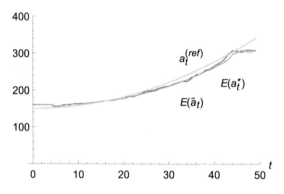

Fig. 10. Average evaluated price paths $E(\tilde{a}_t)$ (blue curve), cf. Algorithm 6.1, compared to optimal price paths $E(a_t^*)$ (orange curve) of the model with full information, cf. Fig. 2, and reference prices $a_t^{(ref)}$ (green curve); Example 6.1, see [9]. (Color figure online)

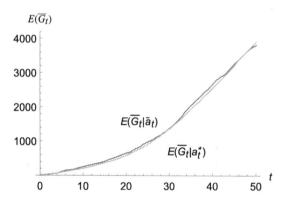

Fig. 11. Average accumulated profits $E(\bar{G}_t|\tilde{a}_t)$ (blue curve) of strategy $\tilde{a}_t$, cf. Algorithm 6.1, compared to average accumulated profits $E(\bar{G}_t|a_t^*)$ (orange curve) of the optimal policy a_t^* of the model with full information, cf. (14); Example 6.1, see [9]. (Color figure online)

7   Conclusion

We studied a stochastic dynamic finite horizon model for sellers to determine price adjustments in the presence of strategic customers. In contrast to classical myopic setups, we consider customers that (i) recur and (ii) compare the current consumer surplus against an expected future surplus, which is based on anticipated future prices, i.e., reference prices.

We derive a pricing strategy that maximizes expected profits by accounting for time-dependent demand and the number of consumers (that are expected to recur). The latter constitute an additional plannable demand potential which can be exploited by the seller.

Further, our framework is characterized by the average evolution of customers' reservation prices and the predetermined reference prices. We show how our model can be used to derive consistent reference prices by iteratively updating them based on the realized offer prices of simulated evaluations of the seller's strategy. We verify that the path of consistent reference prices and average offer prices does not depend on initial starting values.

Our results show, that the behavior of the customers has a significant impact on prices and expected profits of the seller. We find that recurring customers lead to higher average prices, delayed sales, and most importantly higher profits. In contrast, the presence of forward-looking customers has opposing effects. However, it turns out that the latter impact is overcompensated by the effect of recurring customers.

Regarding the expected consumer surplus, we find that a recurring customer behavior leads is not beneficial for consumers as the seller can better anticipate and exploit future demand. Instead, a forward-looking customer behavior that includes anticipations of the seller's offer prices is advantageous for customers as it effectively increases their surplus. In addition, we find that the impact of recurring customers is overcompensated by those of anticipating customers.

Using a Hidden Markov model (HMM), we also consider the case in which the seller cannot distinguish between new customers and returning customers. By comparing solutions of the extended model and the basic model that exploits full information, we verified that the HMM model provides close to optimal solutions.

Appendix

Table 5. Notation table.

T	time horizon/number of periods
t	time/period
N	initial inventory level
X_t	random number of items to sell in t
K_t	random number of waiting customers
G_t	random future profits from t on
c	shipping costs
δ	discount factor
R_t	individual reservation price in t
$\bar{R}_t$	average reservation price in t
$a_t^{(ref)}$	reference price in t
A	set of admissible prices
n	current inventory level
j	number of new customers
i	number of buying customers
k	number of waiting customers
$V_t(n,k)$	value function
a	offer price
CS_t	individual consumer surplus in t
$p_t^{(buy)}(a)$	purchase probability for price a
$p_t^{(new)}(j)$	probability for j new customers
$p_t^{(wait)}(k)$	probability for k waiting customers
$p_t^{(demand)}(i)$	probability for i sales in period t
$a_t^*(n,k)$	optimal prices (full information)
$\pi_t^{(k)}$	beliefs for k waiting customers
$\tilde{a}_t(n)$	heuristic prices (HMM model)
η_t	share of waiting customers
γ_t	share of anticipating customers
$\bar{G}_t$	random accumulated profits up to t
ECS_t	average consumer surplus per sale in t
$ECS_t^{(cum)}$	average accumulated consumer surplus until time t

References

1. Chen, M., Chen, Z.L.: Recent developments in dynamic pricing research: multiple products, competition, and limited demand information. Prod. Oper. Manag. **24**(5), 704–731 (2015)
2. Chenavaz, R., Paraschiv, C.: Dynamic pricing for inventories with reference price effects. Econ. E-J. **12**(64), 1–16 (2018)
3. Coase, R.H.: Durability and monopoly. J. Law Econ. **15**(1), 143–149 (1972)
4. Goensch, J., Klein, R., Neugebauer, M., Steinhardt, C.: Dynamic pricing with strategic customers. J. Bus. Econ. **83**(5), 505–549 (2013)
5. Levin, Y., McGill, J., Nediak, M.: Dynamic pricing in the presence of strategic consumers and oligopolistic competition. Manag. Sci. **55**(1), 32–46 (2009)
6. Liu, Q., Zhang, D.: Dynamic pricing competition with strategic customers under vertical product differentiation. Manag. Sci. **59**(1), 84–101 (2013)
7. Phillips, R.L.: Pricing and Revenue Optimization. Stanford University Press, Palo Alto (2005)
8. Popescu, I., Wu, Y.: Dynamic pricing strategies with reference effects. Oper. Res. **55**(3), 413–429 (2007)
9. Schlosser, R.: Stochastic dynamic pricing with strategic customers and reference price effects. In: International Conference on Operations Research and Enterprise Systems, pp. 179–188 (2019)
10. Schlosser, R., Boissier, M.: Dealing with the dimensionality curse in dynamic pricing competition: using frequent repricing to compensate imperfect market anticipations. Comput. Oper. Res. **100**, 26–42 (2018)
11. Schlosser, R., Boissier, M.: Dynamic pricing under competition on online marketplaces: a data-driven approach. In: 24th ACM SIGKDD International Conference on Knowledge Discovery and Data Mining 2018 (KDD 2018), pp. 705–714 (2018)
12. Schlosser, R., Richly, K.: Dynamic pricing competition with unobservable inventory levels: a hidden Markov model approach. In: Parlier, G.H., Liberatore, F., Demange, M. (eds.) ICORES 2018. CCIS, vol. 966, pp. 15–36. Springer, Cham (2019). https://doi.org/10.1007/978-3-030-16035-7_2
13. Su, X.: Intertemporal pricing with strategic customer behavior. Manag. Sci. **53**(5), 726–741 (2007)
14. Talluri, K.T., Van Ryzin, G.J.: The Theory and Practice of Revenue Management, vol. 68. Springer, Heidelberg (2004). https://doi.org/10.1007/b139000
15. Wei, M.M., Zhang, F.: Recent research developments of strategic consumer behavior in operations management. Comput. Oper. Res. **93**(5), 166–176 (2018)
16. Wu, L.L.B., Wu, D.: Dynamic pricing and risk analytics under competition and stochastic reference price effects. IEEE Trans. Ind. Inf. **12**(3), 1282–1293 (2016)
17. Wu, S., Liu, Q., Zhang, R.Q.: The reference effects on a retailers dynamic pricing and inventory strategies with strategic consumers. Oper. Res. **63**(6), 1320–1335 (2015)
18. Yeoman, I., McMahon-Beattie, U.: Revenue Management: A Practical Pricing Perspective. Palgrave Macmillan, London (2011). Springer

Tractable Risk Measures for the Selective Scheduling Problem with Sequence-Dependent Setup Times

M. E. Bruni[✉] and S. Khodaparasti

Department of Mechanical, Energy and Management Engineering,
University of Calabria, Rende, Italy
{mariaelena.bruni,sara.khodaparasti}@unical.it

Abstract. Quantifying and minimizing the risk is a basic problem faced in a wide range of applications. Once the risk is explicitly quantified by a risk measure, the crucial and ambitious goal is to obtain risk-averse solutions, given the computational hurdle typically associated with optimization problems under risk. This is especially true for many difficult combinatorial problems, and notably for scheduling problems. This paper aims to present a few tractable risk measures for the selective scheduling problem with parallel identical machines and sequence-dependent setup times. We indicate how deterministic reformulations can be obtained when the distributional information is limited to first and second-order moment information for a broad class of risk measures. We propose an efficient heuristic for addressing the computational difficulty of the resulting models and we showcase the practical applicability of the proposed approach providing computational evidence on a set of benchmark instances.

Keywords: Machine scheduling · Risk measure · Heuristic

1 Introduction

Modeling and optimization of risk have been objects of intensive research in the last 15 years [14]. The major developments concern recent advances associated with measurement and control of risks via the formalism of risk measures, and their relation to mathematical programming models. A risk measure is a real-valued functional that quantifies the degree of risk involved in a random outcome. Several papers have specified general properties for a suitable risk measure. The most important set of these properties defines the family of coherent risk measures, which have four basic properties: translation invariance, monotonicity, subadditivity, and positive homogeneity. The majority of risk measures introduced in the literature lacks some of the axiomatic properties required for coherency. For example, the classical Value-at-Risk (VaR) measure, which reports the risk level of a random loss by calculating a quantile of its distribution, is not coherent.

© Springer Nature Switzerland AG 2020
G. H. Parlier et al. (Eds.): ICORES 2019, CCIS 1162, pp. 70–84, 2020.
https://doi.org/10.1007/978-3-030-37584-3_4

An other measure of risk, the Conditional Value-at-Risk (CVaR), has emerged as the most popular coherent alternative to remedy the shortcomings of the VaR. It calculates the average loss exceeding the VaR. The problem of implementing this measure, as any other distribution-based risk measure, is that, in practice, the exact form of the distribution function is lacking and only sample data are available for estimating the distribution. This issue has motivated the development of worst-case risk measures, where the goal is to determine the worst-possible risk level over a set of candidate distributions within a given family sharing certain known moments or known structural properties. The Worst-Case Value-at-Risk was first studied by El Ghaoui et al. [12], who considered a set of candidate distributions described by the first two moments. One of the most remarkable results is the closed-form solution for the Worst-Case Value-at-Risk which provides useful insight into how it can be minimized. Given the elegancy of the closed form, which remarkably resembles the weighted mean-standard deviation, it is natural to wonder if similar results can be also found for alternative risk measures. Recently, Li [15] showed that similar closed-form solutions also exist for the general coherent class of spectral risk measures, of which the CVaR is a special case. The main goal of this paper is to introduce and efficiently solve the selective scheduling problem with sequence-dependent setup times under this wide class of risk measures.

The issue of the uncertainty in scheduling problems has attracted the attention of researchers and practitioners in the last years [6,8,9,11,19]. Most contributions in the literature focus on models in which the uncertainty is handled by replacing the random parameter with an estimated value and the choice under uncertainty is characterized by the maximization or minimization of expected values (see, for instance, [16,19], and the references therein). This framework implements the decision maker's indifference to risk when making a decision. The expected criterion has serious pitfalls in practical decision making problems since it completely disregard the risk associated to a given decision. A few contributions, in the extant machine scheduling literature, deal with the issue of risk. Atakan et al. in [5], proposed a risk-averse approach to tackle the uncertainty of processing times in a single-machine scheduling problem through probabilistic constraints imposed on traditional performance measures such as the total weighted completion time or the total weighted tardiness. The model is solved using a scenario decomposition approach providing promising results. Sarin et al. [18] presented a scenario-based mixed-integer program formulation to minimize CVaR for general scheduling problems. They developed an integer L-shaped algorithm as well as a dynamic programming-based heuristic approach. Chang et al. [10] adopted the distributionally robust approach to sequence a set of jobs with uncertain processing times on a single machine. They formulated the problem as an integer second-order cone programming which is solved using some approximation algorithms. Recently in [7], a new scheduling problem with parallel identical machines and sequence-dependent setup times has been proposed. The model aims at minimizing the total completion time while the total revenue gained by the processed jobs satisfies the manufacturer's threshold. The authors proposed a risk-averse approach to tackle the uncertainty under the assumption that the uncertain setup times and processing times are defined by random vectors with known mean and variance and that the distributional set is a semi-infinite support set. This paper, along this line, further extends the model, considering a broader class of risk measures. To efficiently solve the resulting optimization problem, we propose an efficient heuristic mainly based on a Variable Neighbourhood Descent (VND) mechanism, whose performance is shown through extensive computational results. The rest of the paper is organized as follows. In Sect. 2, the problem statement as well as the proposed mathematical formulation are provided. In Sect. 3, a heuristic approach is proposed. Section 4 presents the computational experiments carried out on a set of benchmark problems. Finally, Sect. 5 concludes the paper.

2   Problem Statement

A typical parallel machine scheduling problem deals with the sequencing and scheduling of a set of jobs to be processed on a set of identical machines where each job is processed by only one machine and each machine can process only one job at a time. A processing time is assigned to each job and setup times might be present, which, generally, are sequence-dependent. The aim is to find a feasible schedule optimizing a time-related performance criterion such as the total completion time. A new class of problems with order acceptance and/or rejection has emerged in the last years, overcoming the assumption that all the jobs have to be processed [17]. In practical settings, in fact, it is impossible to accept all the orders (jobs) due to the resource limitations and to the trade-off between the revenue gained by processing a job and the increase in the total completion time. Following this research stream, in [7] a new selective scheduling problem with profits and sequence-dependent setup times has been proposed in the multi-machine environment with the objective of minimizing the total completion time. In particular, let $V = \{0, 1, \ldots, i, \ldots, j, \ldots, n\}$ be the set of potential jobs. Note that $0 \in V$ is a dummy starting job. Each potential job i requires a processing time p_i and guarantees a profit ψ_i; obviously, $p_i = \psi_i = 0$. A setup time s_{ij} is required to prepare the machine after processing job i and before starting job j. The aim is to select a subset of jobs to be scheduled on k identical machines such that the total profit is above a minimum required level β determined by the manufacturer, while the total completion time is minimized.

To present the mathematical model, let define $G = (V, E)$ as directed graph defined over the set V, and $E = \{(i, j) \in V \times \bar{V} | i \neq j\}$ ($\bar{V} = V - \{0\}$) as the set of arcs representing every possible job sequence. We also consider a level set $L = \{1, \ldots, r, \ldots, N\}$, where $N = |\bar{V}| - k + 1$ is un upper bound on the number of jobs that each machine can process. Each level r denotes the inverse position of a job over each machine (level 1 denotes the last position). Two binary decisions variables can be then defined: x_i^r, which takes value 1 *iff* job i is processed at level r and otherwise is set to 0 and y_{ij}^r which takes value 1 *iff* job j is processed at level r after job i (which was processed at level $r + 1$) and otherwise is set to 0.

With the above notation, the model can be formalized as follows.

$$\min \sum_{r=1}^{N} \sum_{j \in \bar{V}} r \left(s_{0j} + p_j \right) y_{0j}^r + \sum_{r=1}^{N-1} \sum_{i \in \bar{V}} \sum_{\substack{j \in \bar{V} \\ j \neq i}} r \left(s_{ij} + p_j \right) y_{ij}^r \tag{1}$$

$$\sum_{r=1}^{N} x_i^r \leq 1 \quad i \in \bar{V} \tag{2}$$

$$\sum_{r=1}^{N} \sum_{j \in \bar{V}} y_{0j}^r = k \tag{3}$$

$$\sum_{i \in \bar{V}} x_i^1 = k \tag{4}$$

$$\sum_{\substack{j \in \bar{V} \\ j \neq i}} y_{ij}^r = x_i^{r+1} \quad i \in \bar{V}, \ r = 1, 2, \ldots, N-1 \tag{5}$$

$$y_{0j}^r + \sum_{\substack{i \in \bar{V} \\ i \neq j}} y_{ij}^r = x_j^r \quad j \in \bar{V}, \ r = 1, 2, \ldots, N-1 \tag{6}$$

$$y_{0j}^N = x_j^N \quad j \in \bar{V} \tag{7}$$

$$\sum_{i \in \bar{V}} \sum_{r=1}^{N} \psi_i x_i^r \geq \beta \sum_{i \in \bar{V}} \psi_i \tag{8}$$

$$x_i^r \in \{0, 1\} \quad i \in \bar{V}, \ r = 1, 2, \ldots, N \tag{9}$$

$$y_{ij}^N = 0 \quad i \in \bar{V}, \ j \in \bar{V}, \ i \neq j \tag{10}$$

$$y_{ij}^r \in \{0, 1\} \quad i \in V, \ j \in \bar{V}, \ i \neq j \ r = 1, 2, \ldots, N \tag{11}$$

The objective function (1) minimizes the total completion times for all the processed jobs, including the setup and the processing times. The set of constraints (2) ensures that each job is processed at most once. Constraints (3) and (4) require the process of exactly k jobs at the start and at the end of process, respectively. The set of constraints in (5)–(7) are related to the connectivity of the problem. Constraints (5) require that any job i processed at the upper level $r+1$ should be followed by exactly one task (let say j) by traversing arc (i, j) at the lower level r. The set of constraints in (6) impose that any job j processed at level r should be linked to exactly one recently processed task (let say i) by traversing arc (i, j) or linked directly to the dummy job by traversing arc $(0, j)$ at the same level. Constraints (7) guarantee that the first processed job, at the highest level N, should be the processed just after the dummy task by traversing arc $(0, j)$. Constraint (8) states that the total profit gained from processing the selected jobs should be above a predefined threshold. Constraints (9)–(11) define the binary nature of variables.

Let assume that the uncertain setup times $\tilde{s}_{ij}$ and processing times $\tilde{p}_j$ are defined by random vectors $\boldsymbol{s}$ and $\boldsymbol{p}$. The total completion time

$$X = \sum_{r=1}^{N} \sum_{j \in \bar{V}} r \left(\tilde{s}_{0j} + \tilde{p}_j \right) y_{0j}^r + \sum_{r=1}^{N-1} \sum_{i \in \bar{V}} \sum_{\substack{j \in \bar{V} \\ j \neq i}} r \left(\tilde{s}_{ij} + \tilde{p}_j \right) y_{ij}^r$$

is itself a random variable governed by a probability distribution function F_X, defined on a given probability space $(\Omega, \mathcal{F}, \mathbb{P})$, where $\mathcal{F}$ is a $\sigma - algebra$ of subsets of Ω.

Formally, a risk measure is a map $\rho : \mathcal{X} -> \mathcal{R}$ that attaches a scalar value to each random variable $X : \Omega -> \mathcal{R}$, whose moment-generating function $M_X(z) = \mathbb{E}(e^{zX})$ exists for all $z \geq 0$. Artzner et al. [4] stated a set of properties that should be desirable for any risk measure. The four axioms they stated are Monotonicity, Translation equivariance, Subadditivity, and Positive Homogeneity. Given two random variables X and Y and a risk function, ρ we can define the properties as follows.

- Monotonicity- A risk measure is monotone, if for all X, $Y : X \leq Y$ $\rho(X) \leq \rho(Y)$, i.e., higher losses mean higher risk
- Translation Equivariance- A risk measure is translation equivariant, if for all X, and scalars $c \in \mathbb{R}$: $\rho(X + c) = \rho(X) + c$, i.e., increasing (or decreasing) the loss increases (decreases) the risk by the same amount
- Subadditivity- A risk measure is subadditive, if for all X, Y :$\rho(X + Y) \leq \rho(X) + \rho(Y)$, i.e., diversification decreases risk
- Positive Homogeneity- A risk measure is positively homogeneous, if for all X, $\lambda \geq 0$: $\rho(\lambda X) = \lambda \rho(X)$, i.e., doubling the size doubles the risk

Any risk measure which satisfies these axioms is said to be coherent. On of the most popular risk measure is the VaR that, at a given confidence level $\alpha \in (0, 1)$ is denoted by VaR_α and can be described as follows

$$VaR_\alpha = \inf_\eta (\eta | F(\eta) \geq \alpha)$$

More formally, the VaR is defined such that the probability that the random variable X is greater than VaR is less than or equal to $1 - \alpha$ (since $F(\eta) = \mathbb{P}(X \leq \eta)$) and is equivalent to the left-continuous inverse of the cumulative distribution function $(F^{-1}(\alpha))$. In other words, the VaR is the smallest value of X if we exclude worse outcomes whose probability is less than $1 - \alpha$. Despite its wide use, VaR is not a coherent risk measure since lacks subadditivity, even if it satisfies all the remaining axioms of monotonicity, translation equivariance and positive homogeneity. Beside this, an important drawback of the VaR measure is that it does not give any information about the expected value of the worst cases.

The CVaR is closely linked to VaR, but provides several distinct advantages. In fact, the CVaR is consistent with the second-degree stochastic dominance and it is coherent in the terminology of [4]. Basically, CVaR is defined as the average of the $\alpha\%$ worst cases weighted with a uniform weight. More formally, the CVaR risk measure at a given confidence level $\alpha \in (0,1)$, denoted by $CVaR_\alpha$. This risk measure quantifies the expected loss of the random variable in the worst $(1 - \alpha)\%$ of cases described as follows:

$$CVaR_\alpha = \mathbb{E}[X|X \geq VaR_\alpha].$$

If F_X is continuous, then we have

$$CVaR_\alpha = \frac{1}{1-\alpha} \int_\alpha^1 VaR_p dp$$

In most real-life applications, the probability distribution is typically unknown and only indirectly observable through historical samples. A remedy for this difficulty is to adopt a distributionally robust approach, assuming that the probability distribution is merely known to belong to an ambiguity set, typically defined as the family $\mathbb{F}$ of all distributions that have known first and second moments. This ambiguity prompts us to investigate the quantification of the risk in this more general setting. Considering the above definitions, the risk criterion reduces to the worst-case CVaR denoted by, $WCVaR_\alpha$, and defined as follows:

$$WCVaR_\alpha = sup_{F \in \mathbb{F}} CVaR_\alpha \tag{12}$$

In this case, solutions are evaluated under the worst-case over all the distributions in the family $\mathbb{F}$ and hence, consistent with the known moments. The resulting WCVaR represents a conservative (that is, pessimistic) approximation for the true (unknown) CVaR.

A generalization of the CVaR is represented by the class of spectral risk measures, first introduced by Acerbi [1] who attempted to generalize CVaR.

A spectral risk measure, denote by SRM_ϕ is a function parameterized by ϕ, a nondecreasing normalized right-continuous integrable probability density function, such that $\phi \geq 0$, and $\int_0^1 \phi(p)dp = 1$. The density function ϕ is also called an risk spectrum. It can be defined as follows

$$SRM_\phi = \int_0^1 \phi(p)F^{-1}(p)dp = \int_0^1 \phi(p)VaR_p dp.$$

Essentially, a spectral risk measure may be viewed as a weighted sum of VaR, where larger weights are assigned to a larger VaR (this is a consequence of the nondecreasing property of the risk spectrum). We can easily see that CVaR is a special case of spectral risk measures, when

$$\phi(p) = \begin{cases} \frac{1}{1-\alpha} & \text{if } p > \alpha \\ 0 & \text{if } p \leq \alpha. \end{cases}$$

and that spectral risk measures satisfy the properties of monotonicity, convexity, translation invariance and coherency. In analogy with the definition of the

WCVaR, we can define the Worst-Case Spectral Risk Measures (WCSRM):

$$WCSRM = sup_{F \in \mathbb{F}} SRM_\phi.$$

As proposed in [15], it can be proved that the WCSRM is tractable in its full generality and admits an elegant closed form expression, similar to the one obtained for the WCVaR.

Theorem 1. *Let* $\mathbb{F}$ *be the set of all probability distributions with mean* μ *and variance* σ. *For any random variable* $X \in \mathbb{R}^+$, *with a distribution function* F *belonging to the distributional set* $\mathbb{F}$, *any worst-case spectral risk measure WCSRM can be evaluated in closed-form as follows:*

$$WCSRM = \mu + \sigma \sqrt{\int_0^1 \phi^2(p)dp - 1}$$

Proof. See [15]. $\square$

Provided that the term $\int_0^1 \phi^2(p)dp$ can be solved offline, this final problem can be solved by a commercial solver. In the case of CVaR, we have $\int_0^1 \phi^2(p)dp = \frac{1}{1-\alpha}$ that gives rise to the well known formula

$$WCVaR = \mu + \sigma \sqrt{\frac{\alpha}{1 - \alpha}}$$

Remark. A similar closed form (assuming Normal distributions) can also be obtained for another well-known risk measure, the Entropic VaR (EVaR), recently introduced in Ahmadi-Javid [2,3]. The EVaR satisfies most desirable properties that have been postulated by the modern risk theory, namely the property of monotonicity, convexity, translation invariance, coherency, and law invariance.

The EVaR of X with confidence level α is defined as follows:

$$EVaR_\alpha := inf_{z>0}\{z \, ln(M_X(z^{-1})/(1-\alpha))\} =$$
$$= inf_{z>0}\{z \, ln\mathbb{E}\left[\exp\left(\tfrac{X}{z}\right)\right] - z \, ln(1-\alpha)\}$$

It can be shown that EVaR is the tightest possible upper bound for VaR and the CVaR, at the same level of confidence which means that EVaR is known to be more risk-averse compared to others. Despite its seeming complexity, also the EVaR can be boiled down to a closed form expression assuming normally distributed random variables. For a normally distributed random variable $X \sim \mathcal{N}(M, \Sigma)$, we can derive a deterministic equivalent for the EVaR. In particular,

$$EVaR_\alpha := M + \Sigma\sqrt{-2ln(1 - \alpha)}.$$

Theorem 1 provides a baseline to present an equivalent mixed integer mathematical model for the model (1)–(11) under risk. In fact, minimizing the WSRM entails solving the following model:

$$\min z = \sum_{r=1}^{N} \sum_{j \in \bar{V}} r \left[\mu(\tilde{s}_{0j}) + \mu(\tilde{p}_j)\right] y_{0j}^r + \sum_{r=1}^{N} \sum_{i \in \bar{V}} \sum_{\substack{j \in \bar{V} \\ j \neq i}} r \left[\mu(\tilde{s}_{ij}) + \mu(\tilde{p}_j)\right] y_{ij}^r + \Gamma \sqrt{b}$$

(13)

s.t.   (2)–(11)

where

$$b = \sum_{r=1}^{N} \sum_{j \in \bar{V}} r^2 \left[\sigma^2(\tilde{s}_{0j}) + \sigma^2(\tilde{p}_j)\right] y_0^r + \sum_{r=1}^{N} \sum_{i \in \bar{V}} \sum_{j \in \bar{V} j \neq i} r^2 \left[\sigma^2(\tilde{s}_{ij}) + \sigma^2(\tilde{p}_j)\right] y_{ij}^r$$

and $\Gamma = \sqrt{\int_0^1 \phi^2(p)dp - 1}$ for a given spectrum ϕ.

The above result provides a unified perspective on solving the problem under different risk measures in closed form, just by modifying the scale factor Γ of the standard deviation accordingly.

This means that one can use existing optimization methods for solving this second order cone programming problem. These problems, beyond having rich theoretical advantages, can be solved very reliably using existing general-purpose optimization packages. However, to achieve more computational efficiency, we have designed a heuristic algorithm, that will be presented in the next Section.

3   The Heuristic Approach

We now focus on the solution procedure for Problem (13), (2)–(11) presenting a heuristic approach exploiting the specific problem structure. Particularly, we propose a Variable Neighborhood Search (VNS) based heuristic. The VNS is a metaheuristic proposed in [13] as a general framework to solve hard problems. It is based the simple idea of switching to different neighborhood structures starting from the simplest neighborhood type. Note that the order of neighborhood structures is fixed. The exploration of the neighborhood can be carried out randomly or deterministically.

The variant of VNS that explores neighborhoods in a deterministic way is called VND. In each VND iteration, a local search procedure through a given neighborhood structure is applied, followed by a neighborhood change function which defines the neighborhood structure that will be examined in the next iteration. The VND stops when there is no improvement with respect to any of the considered neighborhood structures. Thus, the solution is a local optimum with respect to all neighborhood structure. To overcome this disadvantage, some authors augmented the VND with an intensification mechanism such as a local search. In our algorithm, after the VND, we run a local search to deeply explore the current portion of the solution space. The general structure of the proposed heuristic is sketched in Algorithm 1.

Algorithm 1. The VND heuristic.

---

1 **Input:** s_{ini}
2 **Initialization:** $s \leftarrow s_{ini}$, $s* \leftarrow s_{ini}$, $s' \leftarrow null$
3 **repeat**
4 $\quad$ $\nu \leftarrow 1$
5 $\quad$ **if** $(z(s) < z(s^*))$ **then**
6 $\quad\quad$ | $\quad$ $s^* \leftarrow s$
7 $\quad$ **end**
8 $\quad$ **repeat**
9 $\quad\quad$ | $\quad$ $s' \leftarrow Neighborhood\ search(s, \nu)$
10 $\quad\quad$ | $\quad$ **if** $(z(s') < z(s))$ **then**
11 $\quad\quad$ | $\quad\quad$ | $\quad$ $s \leftarrow s'$
12 $\quad\quad$ | $\quad$ **else**
13 $\quad\quad$ | $\quad\quad$ | $\quad$ $\nu \leftarrow \nu + 1$
14 $\quad\quad$ | $\quad$ **end**
15 $\quad$ **until** $(\nu > N_{max})$;
16 **until** $(z(s) < z(s^*))$;
17 *Local search*
18 **return** $s*$

The *Neighborhood search*(s, ν) heuristic consists in finding an improving direction within a neighbourhood ν centered at the solution s, and moving to the minimum. This strategy is also referred to as the best improvement (all the neighbours in the selected neighbourhood are evaluated and the most improving is executed). Five different neighborhood structures classified as intra-machine and inter-machines move operators are proposed. To be more precise, we consider the following neighborhoods:

1. Selective neighborhoods, where we allow the possibility of modifying the set of selected jobs:
   - *Delete* move.
     This move simply deletes a job from the sequence of jobs.
   - *Insert* move.
     It inserts a new job into a machine schedule.
   - *Replace* move.
     This move replaces a scheduled job with a non-scheduled one.
2. Inter–machines neighborhoods:
   - *Exchange* move.
     It exchanges a pair of jobs between two different machines.
   - *Relocation* move.
     It deletes a job from a path and inserts it to another machine.

The algorithm is initialized with an initial solution s_{ini} generated in a greedy fashion. For each job the completion time $CT_{(.)}$ is evaluated considering the job as the first in the schedule. In particular, the completion time of job $i \in \bar{V}$ is calculated as $CT_i = \mu(\tilde{s}_{0i}) + \mu(\tilde{p}_i) + \sqrt{\sigma^2(\tilde{s}_{0i}) + \sigma^2(\tilde{p}_i)}$. Then, k different jobs with the lowest completion time are selected and assigned to k different machines. If this initial solution is feasible with respect to the constraint (8),

Table 1. Results for the instances with $n = 10$ and $k = 1$.

Instance	$\alpha = 0.1$			$\alpha = 0.5$			$\alpha = 0.9$		
	Gap%	CPU_{VND}	CPU_{SCIP}	Gap%	CPU_{VND}	CPU_{SCIP}	Gap%	CPU_{VND}	CPU_{SCIP}
Tao1R1	9.28	0.01	4.41	5.61	0.01	4.25	5.28	0.01	4.99
Tao1R3	9.71	0.01	4.66	6.54	0.02	4.56	8.75	0.01	5.16
Tao1R5	7.58	0.01	5.05	2.97	0.01	4.59	0.82	0.01	4.92
Tao1R7	4.67	0.01	4.83	4.09	0.01	4.62	2.55	0.01	4.86
Tao1R9	5.11	0.01	5.29	6.18	0.01	5.02	7.56	0.01	5.56
Tao3R1	3.92	0.01	5.31	1.86	0.01	4.94	3.22	0.01	5
Tao3R3	4.11	0.01	5.03	2.52	0.01	4.77	1.85	0.01	4.59
Tao3R5	7.6	0.01	5.45	6.3	0.01	5.04	6.42	0.01	5.38
Tao3R7	12.15	0.01	4.54	11.49	0.01	4.49	10.75	0.02	4.89
Tao3R9	5.09	0.01	5.07	4.51	0.01	5.07	4.01	0.01	5.14
Tao5R1	11.35	0.01	4.46	10.62	0.01	5.25	6.68	0.01	4.81
Tao5R3	10.72	0.01	4.88	11.11	0.01	5.39	9.8	0.01	5.06
Tao5R5	9.25	0.01	4.46	9.23	0.01	5.62	14.17	0.01	5.19
Tao5R7	5.03	0.01	4.45	5.42	0.01	4.43	4.68	0.01	4.76
Tao5R9	13.74	0.01	4.82	10.98	0.01	4.76	8.38	0.01	5.14
Tao7R1	2.34	0.01	4.84	1.79	0.01	5.07	1.58	0.01	5.18
Tao7R3	6.49	0.01	4.57	5.53	0.01	4.7	4.53	0.01	4.42
Tao7R5	6.33	0.01	4.69	6.44	0.01	4.88	5.89	0.01	4.99
Tao7R7	7.56	0.01	4.22	4.57	0.01	4.81	3.94	0.01	5
Tao7R9	1.09	0.01	4.17	2.1	0.01	4.84	1.64	0.01	5.18
Tao9R1	4.3	0.02	4.71	4.93	0.01	4.76	5.37	0.01	4.75
Tao9R3	13.25	0.01	4.76	10.83	0.01	4.35	10.68	0.01	4.48
Tao9R5	3.77	0.01	5.34	4.52	0.01	4.67	4.11	0.01	5.13
Tao9R7	5.86	0.01	4.86	4.97	0.01	5.65	3.51	0.01	5.2
Tao9R9	10.19	0.02	5.93	9.57	0.02	5.02	12.18	0.01	5.02
Avg	7.22	0.01	4.83	6.19	0.01	4.86	5.93	0.01	4.99

the procedure is finished; otherwise, it continues by adding one by one the most profitable non-processed jobs to each machine in a balanced fashion, until the feasibility is reached.

The local search includes *Swap*, *2-opt*, and *Or-opt* moves.

- *Swap* move.
  This operator looks for a pair of jobs assigned to the same machine that would lead to an improved solution if swapped. If such a pair exists, the swap is made and the procedure is repeated.
- *2-opt* move.
  Given a sequence of jobs $(1, 2, \ldots, i, i + 1, \ldots, j, j + 1, \ldots)$ assigned to the same machine, it picks a pair of non-adjacent jobs i and j and enforces adjacency between them. The other jobs are then reconnected as follows $(1, 2, \ldots, i, j, \ldots, i + 1, j + 1, \ldots)$.
- *Or-opt* move.
  This move removes a triple of non-adjacent jobs i, j, k from the sequence

Table 2. Results for the instances with $n = 10$ and $k = 2$.

Instance	$\alpha = 0.1$			$\alpha = 0.5$			$\alpha = 0.9$		
	$Gap\%$	CPU_{VND}	CPU_{SCIP}	$Gap\%$	CPU_{VND}	CPU_{SCIP}	$Gap\%$	CPU_{VND}	CPU_{SCIP}
Tao1R1	5.39	0.01	2.18	5.51	0.01	1.63	6.53	0.01	2.27
Tao1R3	9.96	0.01	2.9	10.79	0.01	1.89	10.72	0.01	2.56
Tao1R5	6.67	0.01	2.25	7.53	0.01	1.94	14.68	0.02	2.25
Tao1R7	1.56	0.01	2.16	6.85	0.01	1.84	5.79	0.01	2.05
Tao1R9	9.98	0.01	3.35	6.48	0.02	1.91	8.6	0.02	2.59
Tao3R1	6.3	0.02	2.29	6.96	0.01	1.61	6.09	0.01	2.69
Tao3R3	8.37	0.01	2.26	7.27	0.02	1.81	4.5	0.01	2.24
Tao3R5	10.86	0.01	2.27	10	0.01	1.8	6.84	0.02	2.23
Tao3R7	6.19	0.01	2.21	7.5	0.01	1.53	8.95	0.02	2.14
Tao3R9	3.42	0.01	2.85	3.14	0.01	1.84	4.78	0.02	2.3
Tao5R1	8.13	0.01	2.53	8.17	0.01	1.91	7.9	0.01	2.71
Tao5R3	7.06	0.01	2.79	5.56	0.01	2.05	6.31	0.01	2.74
Tao5R5	12.06	0.01	2.11	12.13	0.01	1.7	9.9	0.01	2.27
Tao5R7	1.54	0.02	2.89	1.52	0.01	1.75	2.24	0.02	2.11
Tao5R9	5.6	0.01	2.37	6.06	0.01	2.23	5.61	0.01	2.77
Tao7R1	10.16	0.01	2.37	5.41	0.01	1.76	3.91	0.01	2.55
Tao7R3	7	0.01	2.09	5.32	0.01	1.75	5.12	0.01	2.09
Tao7R5	7.17	0.01	2.01	7.07	0.01	1.68	11.39	0.01	2.02
Tao7R7	4.09	0.01	2.28	5.5	0.01	2.2	8.07	0.02	2.46
Tao7R9	2.12	0.02	2.52	4.02	0.01	1.7	4.97	0.02	2.23
Tao9R1	7.35	0.01	2.18	4.68	0.01	1.63	4.3	0.01	2.3
Tao9R3	8	0.01	2.03	14.58	0.01	1.41	12.06	0.02	1.95
Tao9R5	2.28	0.01	2.73	2.99	0.01	1.71	3.42	0.02	2.35
Tao9R7	1.73	0.02	2.98	1.46	0.02	1.96	0.71	0.01	2.53
Tao9R9	3.22	0.02	2.44	3.33	0.02	1.91	2.23	0.02	2.47
Avg	6.25	0.01	2.44	6.39	0.01	1.81	6.62	0.01	2.35

$(1, 2, \ldots, i, i + 1, \ldots, j, j + 1, \ldots, k, k + 1, \ldots)$ and reconnects them in such a way that the order of intermediate jobs is preserved $(1, 2, \ldots, i, j+1, \ldots, k, i+1, \ldots, j, k + 1, \ldots)$.

4   Computational Results

To test the heuristic approach, we have selected 750 instances from the benchmark test used for the order acceptance problem in the single machine context ([17]) divided into three groups, each containing 25 data sets each containing 10 instances, with 10, 15, and 25 jobs. The setup times, processing times, and the profits are the same as those reported in the benchmark and the setup time variances $\sigma^2(\tilde{s}_{ij})$ as well as the processing time variances $\sigma^2(\tilde{p}_i)$ were set to $\sigma^2(\tilde{s}_{ij}) = \lceil \zeta_s^2 \rceil$ and $\sigma^2(\tilde{p}_i) = \lceil \zeta_p^2 \rceil$ where ζ_s and ζ_p are random numbers uniformly distributed in intervals $[1, \frac{1}{2}(\min_{i \in V, j \in \bar{V}} \mu(\tilde{s}_{ij}) + \max_{i \in V, j \in \bar{V}} \mu(\tilde{s}_{ij}))]$ and

Table 3. Results for the instances with $n = 15$ and $k = 1$.

Instance	$\alpha = 0.1$			$\alpha = 0.5$			$\alpha = 0.9$		
	$Gap\%$	CPU_{VND}	CPU_{SCIP}	$Gap\%$	CPU_{VND}	CPU_{SCIP}	$Gap\%$	CPU_{VND}	CPU_{SCIP}
Tao1R1	6.86	0.05	28.89	7.68	0.04	29.09	7.28	0.04	28.08
Tao1R3	12.41	0.05	29.27	7.9	0.04	28.67	5.94	0.05	30.41
Tao1R5	5.18	0.05	35.28	4.36	0.05	29.53	6.14	0.05	36.12
Tao1R7	9.24	0.05	31.23	8.07	0.05	33.67	5.86	0.05	35.91
Tao1R9	9.24	0.05	32.34	10.27	0.05	31.01	7.65	0.05	29.16
Tao3R1	6.94	0.04	31.47	6.16	0.05	29.95	7.44	0.05	30.51
Tao3R3	12.38	0.05	32.06	8.57	0.05	27.64	5.7	0.05	29.57
Tao3R5	4.43	0.05	32.51	2.91	0.05	29.07	2.33	0.05	30.85
Tao3R7	7.29	0.05	35.11	4.36	0.05	34.65	2.94	0.05	33.95
Tao3R9	12.11	0.05	32.92	8.56	0.04	29.52	4.47	0.05	29.37
Tao5R1	6.79	0.06	33.93	6.41	0.06	30.99	5.67	0.06	35.1
Tao5R3	9.29	0.04	33.61	9.4	0.04	27.17	11.21	0.04	26.73
Tao5R5	5.72	0.05	35.95	5.4	0.05	31.92	7.95	0.05	33.61
Tao5R7	6.14	0.06	31.19	7.57	0.06	29.8	8.74	0.06	29.13
Tao5R9	8.06	0.04	28.47	6.05	0.04	25.96	5.81	0.04	26.38
Tao7R1	4.66	0.04	31.67	4.67	0.05	29.24	4.06	0.04	28.41
Tao7R3	1.87	0.05	31.66	3.11	0.05	28.44	3.56	0.05	32.74
Tao7R5	8.9	0.05	32.33	7.25	0.06	31.16	6.37	0.06	33.86
Tao7R7	8.02	0.05	28.82	6.69	0.04	28.85	5.18	0.05	28.07
Tao7R9	8.37	0.05	34.41	9.53	0.05	31.72	9.14	0.05	35.09
Tao9R1	7.13	0.04	31.44	7.15	0.05	30.52	5.66	0.05	28.86
Tao9R3	11.81	0.05	29.03	10.62	0.05	26.47	10.55	0.05	23.9
Tao9R5	7.35	0.05	32.55	6.52	0.05	29.21	6	0.05	25.03
Tao9R7	4.68	0.04	31.21	4.34	0.04	33.59	5.61	0.05	34.29
Tao9R9	8.59	0.04	35.2	4.65	0.04	28.27	4.35	0.04	28.8
Avg	7.74	0.05	32.1	6.73	0.05	29.84	6.22	0.05	30.56

$[1, \frac{1}{2}(\min_{i \in V} \mu(\tilde{p}_i) + \max_{i \in V} \mu(\tilde{p}_i))]$, respectively. The value of β is set to 0.6 and $k = 1, 2$. Γ has been set to $\sqrt{\frac{1-\alpha}{\alpha}}$ and three different values of $\alpha = 0.1, 0.5, 0.9$ have been considered, reflecting different risk attitude, form the almost risk neutral to the most risk-averse one. The proposed model was implemented in C++. The experiments were executed on an Intel® Core™ i7 2.90 GHz, with 8.0 GB of RAM memory. The SCIP solver was used to solve the mathematical model with a time limit imposed of 1000 seconds. The results are shown in the Tables 1, 2, 3, 4 and 5. In each Table the name of data set (containing 10 instances) is reported in Column 1. The average gap of the heuristic $Gap\% = \frac{z_{Heu} - z_{SCIP}}{z_{SCIP}} \times 100$ (z_{Heu} and z_{SCIP} are, respectively, the best solution values found by the heuristic and SCIP), and the average CPU for both the SCIP and the VND are reported in the columns with headings CPU_{SCIP} and CPU_{VND}, respectively. The tests for which SCIP was not able to close the optimality gap within the time limit, are highlighted in bold.

Table 4. Results for the instances with $n = 15$ and $k = 2$.

Instance	$\alpha = 0.1$			$\alpha = 0.5$			$\alpha = 0.9$		
	Gap%	CPU_{VND}	CPU_{SCIP}	Gap%	CPU_{VND}	CPU_{SCIP}	Gap%	CPU_{VND}	CPU_{SCIP}
Tao1R1	10.06	0.04	12.56	6.55	0.06	11.3	5.65	0.06	11.64
Tao1R3	8.62	0.05	10.08	7.65	0.05	10.39	7.94	0.05	10.27
Tao1R5	4.74	0.04	12.48	3.65	0.05	11.59	4.81	0.05	11.25
Tao1R7	5.84	0.05	11.74	5.35	0.05	13.63	5.76	0.05	11.67
Tao1R9	11.69	0.05	10.63	10.58	0.04	14.28	8.77	0.04	13.45
Tao3R1	5.96	0.05	12.52	5.91	0.05	10.83	7.98	0.06	10.19
Tao3R3	6.19	0.04	13.5	10.71	0.05	10.34	8.26	0.05	10.31
Tao3R5	4.62	0.05	12.07	3.05	0.04	11.13	1.57	0.04	11.04
Tao3R7	8.15	0.05	10.46	7.52	0.05	12.89	4.48	0.05	12.55
Tao3R9	11.99	0.05	14	9.32	0.05	13.74	7.8	0.04	13.64
Tao5R1	7.22	0.06	15.1	7.81	0.06	13.13	6.14	0.06	13.97
Tao5R3	7.02	0.04	12.53	4.93	0.05	11.23	5.36	0.05	11.33
Tao5R5	5.91	0.05	10.31	4.5	0.05	11.23	6.07	0.04	13.45
Tao5R7	7.39	0.04	10.16	6.15	0.04	11.01	3.6	0.04	11.19
Tao5R9	8.18	0.04	13.98	9.39	0.05	10.86	9.95	0.05	11.22
Tao7R1	6.86	0.04	11.99	7.22	0.05	10.62	4.02	0.05	9.47
Tao7R3	3.73	0.04	10.4	3.95	0.05	11.87	5.48	0.04	11.49
Tao7R5	8.38	0.05	12.07	8.73	0.06	11.45	8.59	0.05	10.98
Tao7R7	4.72	0.04	13.09	3.71	0.04	13.97	3.6	0.04	13.12
Tao7R9	5.83	0.04	12.94	4.85	0.05	16.79	9.56	0.04	11.22
Tao9R1	7.25	0.04	10.67	8.2	0.04	10.96	7.4	0.04	11.72
Tao9R3	6.93	0.04	12.35	7.31	0.05	11.9	6.04	0.05	12.3
Tao9R5	10.48	0.05	10.61	9.74	0.05	11.85	6.46	0.05	14.19
Tao9R7	1.8	0.05	10.08	2.75	0.05	11.04	4.19	0.05	10.9
Tao9R9	7.41	0.04	11.25	7.27	0.05	10.73	2.92	0.04	11.78
Avg	7.08	0.05	11.9	6.67	0.05	11.95	6.1	0.05	11.77

From the analysis of the results of Table 1, we can observe that the heuristic is very fast and the average gap is slightly less than 5%. We notice that the problem can be efficiently solved by SCIP in a few seconds. For $k = 2$, (Table 2) the heuristic has an average gap of 6.25% for $\alpha = 0.1$ and increases to 6.39% and 6.62% with the increase of α to 0.5% and 0.9%, respectively.

In terms of computational time, the heuristic is more efficient than SCIP, which is able to optimally solve all the instances in an average CPU time of a couple of seconds. Tables 3 and 4 show the results for 15 jobs. We observe that the SCIP solution time increases considerably for the one machine case. Nonetheless, the heuristic CPU time is very limited and almost neglectable. We observe that the average gap is around 7%. We should observe that the VND heuristic tends to be trapped in local optima and that the problem under study is quite challenging. The gap decreases when α increases, showing a good performance of the heuristic for risk-averse cases.

When the number of jobs is increased to 25 the case of one machine becomes computationally overwhelming. We report in Table 5 the results for two machines. In this case, not all the instances were solved to optimality, especially

Table 5. Results for the instances with $n = 25$ and $k = 2$.

Instance	$\alpha = 0.1$			$\alpha = 0.5$			$\alpha = 0.9$		
	$Gap\%$	CPU_{VND}	CPU_{SCIP}	$Gap\%$	CPU_{VND}	CPU_{SCIP}	$Gap\%$	CPU_{VND}	CPU_{SCIP}
Tao1R1	9.49	0.25	61.08	9.19	0.24	91.6	7.68	0.25	84.6
Tao1R3	10.7	0.21	138.83	9.61	0.23	151.66	9.45	0.25	**262.53**
Tao1R5	6.53	0.19	84.66	8.13	0.21	94.08	7.43	0.21	309.59
Tao1R7	9.79	0.22	162.2	8.53	0.22	**270.81**	6.61	0.22	**314.76**
Tao1R9	11.41	0.2	146.94	11.63	0.19	163.42	10.2	0.22	**294.09**
Tao3R1	5.77	0.22	194.93	5.69	0.2	**164.86**	5.25	0.23	**423.52**
Tao3R3	8.81	0.23	152.43	8.2	0.23	290.44	5.44	0.25	**480.99**
Tao3R5	4.52	0.22	125.12	5.3	0.2	118.74	5.75	0.17	**279.75**
Tao3R7	8.04	0.19	110.89	9	0.19	155.02	7.82	0.22	213.66
Tao3R9	6.81	0.25	75.07	6.35	0.25	57.21	6.78	0.25	52.1
Tao5R1	8.69	0.21	86.26	7.43	0.25	111.95	8.76	0.3	96.03
Tao5R3	9.88	0.24	72.08	9.78	0.22	62.68	6.81	0.22	134.15
Tao5R5	4.95	0.22	68.91	5.89	0.22	91.08	4.69	0.22	**233.91**
Tao5R7	8.05	0.2	108.93	6.77	0.25	59.54	7.27	0.21	69.64
Tao5R9	7.37	0.2	124.65	6.03	0.2	98.33	6.62	0.18	**467.59**
Tao7R1	7.51	0.25	185.86	9.51	0.23	144.02	7.14	0.22	**283.76**
Tao7R3	7.75	0.2	100	7.76	0.2	133.34	7.78	0.21	**322.55**
Tao7R5	4.86	0.21	63.15	5.2	0.23	88.36	4.86	0.22	**146.03**
Tao7R7	7.26	0.22	98.68	5.19	0.21	128.41	3.78	0.19	**189.01**
Tao7R9	6.97	0.2	163.46	5.27	0.22	134.7	4.87	0.21	**454.13**
Tao9R1	5.89	0.27	119.6	7.56	0.24	94.44	8.06	0.21	**277.01**
Tao9R3	8.46	0.22	214.97	6.07	0.2	182.1	8.07	0.2	**285.03**
Tao9R5	9.02	0.19	109.13	8.27	0.19	83.81	9.21	0.19	233.71
Tao9R7	7.27	0.2	124.89	7.64	0.21	92.03	8.41	0.21	**268.4**
Tao9R9	8.06	0.23	114.26	6.48	0.2	146.64	7.31	0.18	**172.77**
Avg	7.75	0.22	120.28	7.46	0.22	128.37	7.04	0.22	253.97

in the more involved case ($\alpha = 0.9$). The average SCIP CPU in this case, is around 250 s, whilst the heuristic still can solve the problem in less than half a second.

5   Conclusions

In this paper, we introduced the selective scheduling problem with parallel identical machines and sequence-dependent setup times under spectral risk measures. We indicate how deterministic reformulations can be obtained when the distributional information is limited to first and second order moment information and we propose an efficient heuristic for addressing the computational difficulty of the resulting models. We tested the efficiency of the proposed model on a large set of scheduling benchmark instances. Although the proposed algorithm is very fast, developing more efficient algorithms is a future research direction. Other scheduling optimization problems under risk could also be investigated in future studies.

References

1. Spectral measures of risk: a coherent representation of subjective risk aversion. J. Bank. Finance **26**(7), 1505–1518 (2002)
2. Ahmadi-Javid, A.: Addendum to: Entropic value-at-risk: a new coherent risk measure. J. Optim. Theory Appl. **155**(3), 1124–1128 (2012)
3. Ahmadi-Javid, A.: Entropic value-at-risk: a new coherent risk measure. J. Optim. Theory Appl. **155**(3), 1105–1123 (2012)
4. Artzner, P., Delbaen, F., Eber, J.M., Heath, D.: Coherent measures of risk. Math. Finance **9**(3), 203–228 (1999)
5. Atakan, S., Bülbül, K., Noyan, N.: Minimizing value-at-risk in single-machine scheduling. Ann. Oper. Res. **248**(1–2), 25–73 (2017)
6. Bruni, M.E., Beraldi, P., Guerriero, F., Pinto, E.: A scheduling methodology for dealing with uncertainty in construction projects. Eng. Comput. **28**(8), 1064–1078 (2011)
7. Bruni, M.E., Khodaparasti, S., Beraldi, P.: A selective scheduling problem with sequence-dependent setup times: a risk-averse approach, pp. 1–7. SciTePress (2019)
8. Bruni, M.E., Di Puglia Pugliese, L., Beraldi, P., Guerriero, F.: An adjustable robust optimization model for the resource-constrained project scheduling problem with uncertain activity durations. Omega **71**, 66–84 (2017)
9. Bruni, M.E., Di Puglia Pugliese, L., Beraldi, P., Guerriero, F.: A computational study of exact approaches for the adjustable robust resource-constrained project scheduling problem. Comput. Oper. Res. **99**, 178–190 (2018)
10. Chang, Z., Song, S., Zhang, Y., Ding, J.Y., Zhang, R., Chiong, R.: Distributionally robust single machine scheduling with risk aversion. Eur. J. Oper. Res. **256**(1), 261–274 (2017)
11. Emami, S., Moslehi, G., Sabbagh, M.: A benders decomposition approach for order acceptance and scheduling problem: a robust optimization approach. Comput. Appl. Math. **36**(4), 1471–1515 (2017)
12. El Ghaoui, L., Oks, M., Oustry, F.: Worst-case value-at-risk and robust portfolio optimization: a conic programming approach. Oper. Res. **51**(4), 543–556 (2003)
13. Hansen, P., Mladenović, N., Moreno Pérez, J.A.: Variable neighbourhood search: methods and applications. Ann. Oper. Res. **175**(1), 367–407 (2010)
14. Krokhmal, P., Zabarankin, M., Uryasev, S.: Modeling and optimization of risk. Surv. Oper. Res. Manag. Sci. **16**(2), 49–66 (2011)
15. Li, J.Y.M.: Technical note–closed-form solutions for worst-case law invariant risk measures with application to robust portfolio optimization. Oper. Res. **66**(6), 1533–1541 (2018)
16. Niu, S., Song, S., Ding, J.Y., Zhang, Y., Chiong, R.: Distributionally robust single machine scheduling with the total tardiness criterion. Comput. Oper. Res. **101**, 13–28 (2019)
17. Oguz, C., Salman, F.S., Yalçın, Z.B., et al.: Order acceptance and scheduling decisions in make-to-order systems. Int. J. Prod. Econ. **125**(1), 200–211 (2010)
18. Sarin, S.C., Sherali, H.D., Liao, L.: Minimizing conditional-value-at-risk for stochastic scheduling problems. J. Sched. **17**(1), 5–15 (2014)
19. Xu, L., Wang, Q., Huang, S.: Dynamic order acceptance and scheduling problem with sequence-dependent setup time. Int. J. Prod. Res. **53**(19), 5797–5808 (2015)

Applications

Modelling Alternative Staffing Scenarios for Workforce Rejuvenation

Etienne Vincent[(✉)] ⓘ and Peter Boyd ⓘ

Director General Military Personnel Research and Analysis,
Department of National Defence, 101 Colonel by Dr., Ottawa,
ON K1A 0K2, Canada
{Etienne.Vincent,Peter.Boyd2}@forces.gc.ca

Abstract. This paper presents methods for predicting the effect of staffing policies on the age distribution of employees in a Public Service workforce. It describes deterministic models based on the application of rates of personnel flows to workforce segments. The models are refinements of previously published work where attrition among recruits had not been considered. A first model works by solving a system of linear equations describing personnel flows to obtain the workforce's age profile at equilibrium. A second model determines the age profile trajectory by iterating over successive future years. Models are compared and validated with historical data. Furthermore, we investigate different measurements of past attrition rates for use with the models.

Keywords: Personnel modelling · Human resources planning · Workforce analytics · Workforce demographics · Workforce rejuvenation · Attrition rate

1 Background

This paper improves and expands work presented at the 8th International Conference on Operations Research and Enterprise Systems [1]. The original work was initiated in response to a request from the Science and Technology branch of the Canadian Department of National Defence (DND). The request was for analytical support toward the achievement of rejuvenation and employment equity representation objectives among departmental Defence Scientists (DSs). The Defence Scientific Service is an Occupational Classification of the federal Public Service that delivers Defence and Security scientific solutions and advice.

Director General Military Personnel Research and Analysis, an organization that routinely conducts workforce analytics and personnel modelling studies for the Canadian Armed Forces and DND, undertook work on this request and completed the original study in 2017. Results from the study have since influenced strategic hiring policies.

1.1 Analytical Problem

Figure 1 shows the age distribution of DSs as of March 31st 2019. Although it has shifted slightly since the original study in 2017, it still displays the pattern that worried

© Crown 2020
G. H. Parlier et al. (Eds.): ICORES 2019, CCIS 1162, pp. 87–108, 2020.
https://doi.org/10.1007/978-3-030-37584-3_5

Human Resources (HR) planners then. Federal public servants become eligible to retire with an immediate annuity at the age of 55 when they have 30 or more years of service, or otherwise at the age of 60[1]. Some choose to work longer, especially while still accruing pensionable years of service (up to 35), which increases the amount of the annuity. Nevertheless, public servants retire at an increasing frequency after reaching 55. The high proportion of DSs who are approaching or already exceed the age of 55 was the concern of HR planners. A spike in retirements could result in a loss of important expertise, whereas a more gradual stream of departures for retirement increases the chance that the expertise is transferred and maintained.

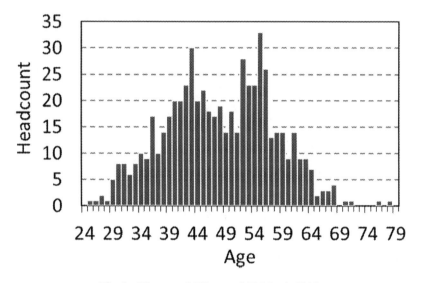

Fig. 1. The age of DSs, as of 31 March 2019.

Figure 2 shows how the mean age of DSs has increased over recent years. This further justified the request from HR planners to study this workforce. There are a few contributing factors to the aging of the DSs workforce. First, some employees have chosen to retire later. This is expected, since the age for pension eligibility was increased in 2013, however this is likely a minor factor as few of the post-2013 hires have yet reached retirement age. Another reason is that the age at which employees are initially hired has tended to increase, in particular due to the fact that increasing proportions of DSs arrive with a Ph.D. Finally, in recent years, the DS workforce has shrunk from a high of 743 DSs in 2010 to 590 in March 2019. As a consequence, continuing employees are getting older, while the departing employees have often not been replaced by recruits who are typically younger.

---

[1] In 2013, immediate annuity eligibility was increased to 60 with 30 years of service, or otherwise 65, but the vast majority of DSs eligible to retire at the time of writing were grandfathered into the pre-2013 rules.

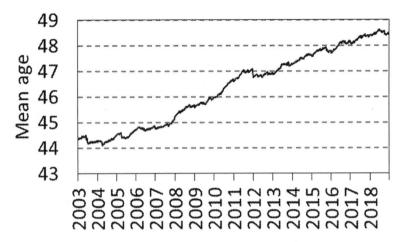

Fig. 2. The mean age of DSs over recent years.

The original rejuvenation study was requested to address the observed ageing of the DS workforce. One of the goals of that study was to identify policies that would result, over time, in a more balanced age distribution. To do this, the study predicted what future age distributions would be, if no corrective action was taken, and then compared that to the range of predicted outcomes under alternative staffing policies.

A related study of the DS workforce was described in [2], but that study focused on career progression, rather than rejuvenation. Past forecasts for other classifications of DND employees have often been based on Discrete Event Simulations (e.g. [3, 4]). Here we present deterministic models based on personnel flows applied to the entire workforce.

1.2   Personnel Data

Since the publication of our original study, we have gained access to new and significantly improved personnel data. This revised and expanded paper benefits from this new data in a number of ways.

Data for the original study was limited to annual workforce snapshots – tables where each row represents an employee, as of the end of each fiscal year. New hires had been identified as those present in the table one year, who had not been there the previous year. Similarly, departing employees were those present one year, but gone the next.

Annual snapshots do not provide information on those recruits who depart before the end of the year when they were hired. We therefore refer to the counts of hires obtained from annual snapshots as *net intake*, as opposed to *gross intake*, which comprises all hires, including those who join and subsequently leave in the same year. Because only annual data was previously available, the models described in our original study were developed to work with *net intake* data.

Annual snapshots also suffered from having been recorded soon after the end of each past fiscal year, and therefore did not often account for subsequently recorded events (e.g. terminations that were effective before the end of the fiscal year, but that had been recorded in the HR system after).

In this work, we had access to historical transaction data – daily records of all past workforce changes, including hires and departures. This allows us to determine accurate headcounts at any point in the past, and to count gross hires and departures over any past periods. Access to finer (daily vs annual) resolution gives us more options for measuring past attrition rates. More specifically, *gross intake* data allows us to develop more complete models of the workforce's evolution, which accounts for the attrition behaviour of new hires. The new data also covers more years, enabling further validation of our models over a large breadth of historical data.

2   Age at Hiring

The Defence Scientific Service Classification has several levels, numbered from 1 to 8. The level of a DS corresponds to their state of career progression, and is tied to a pay scale.

New hires are assigned a level according to an assessment of their education and prior work experience. Figure 3 shows hiring counts, by level and age, between 1 April 2014 and 31 March 2019.

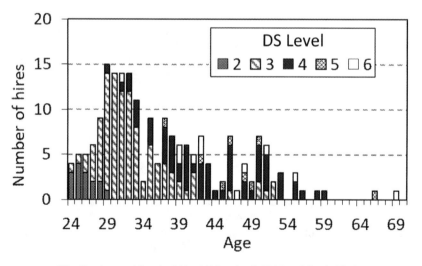

Fig. 3. Age and level of hired DSs, April 2014 to March 2019.

New employees are hired on different dates throughout the year. To facilitate subsequent analysis, we will be tracking age at the time of hire as the age of the employee at the start of the fiscal year in which they were hired (1 April). For example, an employee hired in June, at the age of 50, and with a birthday in May, will be recorded as having been hired at the age of 49.

It is seen, in Fig. 3, that the level assigned to recruits tends to increase with their age. This is because many older hires have acquired professional and academic experience warranting a higher level upon becoming a DS.

Public service staffing competitions are always aimed at specific levels. Prospective employees will only be hired through competitions that target the level that is commensurate with their previously acquired experience. Competitions targeting lower levels bring in less experienced and thus often younger recruits. We note that age discrimination is prohibited when hiring, however there is a correlation between age and level at hire.

Thus, our study of DSs proceeds by forecasting the effect on the eventual workforce age profiles of different apportionments of levels among the positions open to recruits. From April 2014 to March 2019, approximately 8% of the recruits were hired at the second level, 53% at level 3, 30% at level 4, 4% at level 5 and 6% at level 6 (which does not add to 100% due to rounding). At the same time, the mean age at hiring was 26 at level 2, 33 at level 3, 44 at level 4, and 47 at levels 5 and 6. Therefore shifts of the apportionment of levels toward the junior levels would tend to lower the average hiring age. Table 1 shows the six level apportionment scenarios that were investigated in this work. These scenarios had been selected in consultation with the previous study's sponsor [1].

Table 1. The hiring scenarios investigated in this paper.

Scenario	A	B	C	D	E	Current
Level 2	50%	25%	–	20%	20%	8%
Level 3	50%	75%	100%	50%	40%	53%
Level 4	–	–	–	30%	30%	30%
Level 5	–	–	–	–	10%	4%
Level 6	–	–	–	–	–	6%
Mean age	29.8	31.5	33.2	35.0	36.6	37.4

The scenario denoted as *current* repeats the distribution observed in Fig. 3. Scenario A, with 50% level 2 and 50% level 3 hires was thought to be the most extreme hiring regime that would be feasible (Ph.D. graduates automatically start at level 3, and were seen by the study sponsors as an unavoidable recruitment pool). The other scenarios were selected as plausible regimes that would preferentially lower age distribution of new hires to varying degrees. The bottom row of Table 1 also shows the mean hire age that would result from these scenarios, assuming that age distributions for recruits at each level remain unchanged from those observed in recent years.

3 Attrition Rates

Attrition is when employees leave the workforce for any reason. The age at which employees depart impacts the workforce's age distribution, and is therefore critical to our modelling. Attrition is often modelled as a stochastic push process [5]. That is,

departures are modelled as occurring spontaneously according to established probability distributions. Historical attrition rates are usually taken as the model attrition
probabilities, that is, the proportion of all employees that departed in the past is taken as
the probability that any given employee will depart in the future. This section presents
four alternatives for deriving of historical attrition rates. In Sect. 5.3, these four measures will be compared in terms of the accuracy of the models that they yield.

3.1 General Formula for the Attrition Rate

Figure 4 depicts a workforce's headcount over time. At time 0, there are w_0 employees
in this workforce. Employees depart over time, this attrition reduces the headcount to
w_1 at time t_1. However, new employees are also recruited. In Fig. 4, each of the time
steps shown (except times 0 and t_n) correspond to the times when employees were
hired, which we will call *intake events*. The number of recruits associated with each
intake event are r_1, r_2, r_3, and so on, whereas the headcounts immediately before each
intake event are w_1, w_2, w_3, etc.

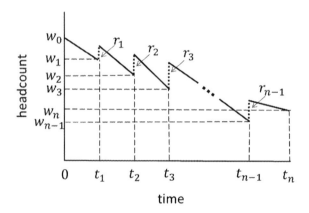

Fig. 4. The headcount of an illustrative workforce over time.

The attrition rate for the workforce shown in Fig. 4 should describe the rate at
which the headcount goes from w_0 to w_n. The complication is in accounting for the
recruitment, which is occurring at the same time.

Over any sub-period without intake, obtaining the attrition rate is straightforward.
For example, over $[0, t_1]$, w_1/w_0 of the employees are retained, and so the attrition rate
is $1 - w_1/w_0$. More generally,

$$A_{[t_i, t_{i+1}]} = 1 - \frac{w_{i+1}}{w_i + r_i} \tag{1}$$

over intake-free sub-periods. Over successive time periods, proportional discount rates, such as attrition rates, compound through multiplication. This means that

$$1 - A_{[t_a,t_c]} = \left(1 - A_{[t_a,t_b]}\right) \times \left(1 - A_{[t_b,t_c]}\right) \tag{2}$$

for any $0 \le t_a \le t_b \le t_c \le t_n$.

Combining Eqs. (1) and (2), we get

$$1 - A_{[0,t_n]} = \frac{w_1}{w_0} \times \frac{w_2}{w_1 + r_1} \times \cdots \times \frac{w_n}{w_{n-1} + r_{n-1}} = \frac{w_n}{w_0} \prod_{i=1}^{n-1} \frac{w_i}{w_i + r_i} \tag{3}$$

In [6], we called Eq. (3) the *general formula* for the attrition rate, and advocate for its use in personnel attrition reporting.

Another field that is also centered on proportional growth and discount rates is Investment Performance Reporting. In that field, the accuracy and defensibility of reports is paramount, the Global Investment Performance Standards [7] mandates the use of Eq. (3) (or close approximations) for reporting the performance achieved by fund managers. In that context, Eq. (3) is called the *time-weighted rate of return*.

Usually, attrition rates are expressed as applying to standard duration periods, most commonly as annual rates. On the other hand, Eq. (3) is for an arbitrary period $[0, t_n]$. If time is measured in years, attrition rates can be annualized as $\bar{A}$, using

$$1 - \bar{A}_{[0,t_n]} = \left(1 - A_{[0,t_n]}\right)^{\frac{1}{t_n}} \tag{4}$$

3.2   Internal Rate of Attrition

From Eq. (3), we can obtain

$$w_n = \left(1 - A_{[0,t_n]}\right) \cdot w_0 + \sum_i \left[\left(1 - A_{[t_i,t_n]}\right) \cdot r_i\right] \tag{5}$$

Intuitively, Eq. (5) determines a future headcount (w_n), as the result of reducing the current headcount (w_0) according to the attrition rate over $[0, t_n]$, while adding the recruits from each intake event (r_i) reduced according to the attrition rates over the remaining periods $[t_i, t_n]$. It seems that we could forecast future headcounts from Eq. (5), but for the fact that it would require knowing what attrition rates will be over all future sub-periods $[t_i, t_n]$. Additional assumptions are needed to obtain a forecasting formula that will be usable in practice.

A natural forecasting assumption is that the attrition rate over future sub-periods will be consistent. That is, the same *annual* attrition rate applies to all periods. In equation form,

$$1 - A_{[t_i,t_n]} = (1 - A')^{t_n - t_i} \tag{6}$$

where A' is the annual attrition rate that applies to the entire forecast, which we will call the *internal rate of attrition*. Substituting Eq. (6) into Eq. (5), we get the internal rate of attrition formula:

$$w_n = (1 - A')^{t_n} \cdot w_0 + \sum_i \left[(1 - A')^{t_n - t_i} \cdot r_i \right] \tag{7}$$

We now have a usable formula for forward-projecting an attrition rate. Equation (7) can also be used to compute an internal rate of attrition from historical data (w_0, w_n, and a series of r_i). To do this, A' must be obtained numerically.

In the same way that the general formula for the attrition rate is analogous to the time-weighted rate of return used in Investment Performance Reporting, the internal rate of attrition is analogous to the 'internal rate of return', a widely used quantity in Finance.

Both the general formula for the attrition rate and the internal rate of attrition were based on recruit counts from the sequence of all intake events. In practice, we can rely on daily resolution data: all recruits show up at the beginning of the workday, and all departures are effective at the end of a workday. Then, the product in the general formula and the sum in the internal rate formula can be taken over all days of the considered time interval.

3.3   Half-Intake Attrition Rate Approximation

Using the general formula for the attrition rate or internal rate of attrition formula requires a catalogue of all successive intake events. However, in many cases, only periodic (e.g. annual or monthly) totals are available, or when using the internal rate to predict future attrition, only annual intake projections are provided. Approximations of these attrition rate formulas are needed for such situations.

Approximations of the formulas can be obtained by making an assumption about how intake totals are distributed across the period. For example, it could be assumed that annual intake is uniformly distributed across the year, or that it follows a seasonal pattern. In [6] we considered several attrition rate approximation formulas derived by distributing intake differently in either the general or internal rate formulas, and compared them empirically on historical Canadian Armed Forces personnel data. The preferred approximation formula from an annual intake total emerged as the half-intake formula:

$$A''_{[0,1]} = \frac{w_0 + r - w_1}{w_0 + \frac{r}{2}} \tag{8}$$

Intuitively, the numerator of Eq. (8) is the number of employees who departed over the year, and the denominator approximates the number of employees eligible to depart (the headcount at year start plus half of the recruits, whose intake is distributed evenly throughout the year). Equation (8) can be derived from either the general formula or internal rate formula, by evenly distributing an annual intake (r) between the first and last day of the year.

In the area of workforce modelling, the half-intake formula was initially proposed by Okazawa [8]. In Investment Performance Management, it had been popularized earlier as the Simple Dietz method [9], occasionally used as an approximation of the Time-Weighted Rate of Return.

When annual headcounts and intake totals are to be used to measure attrition over a multi-year period, the preferred approximation of the general formula investigated in [6] proved to be geometrically linked annual half-intake approximations. The half-intake formula is applied annually, and the resulting annual rates are combined by multiplying their complements. This works out to

$$1 - A''_{[0,n]} = \prod_{i=1}^{n} \frac{w_i - \frac{r_i}{2}}{w_{i-1} + \frac{r_i}{2}} \tag{9}$$

where n is the number of years, w_i is the headcount at the end of the i th year, and r_i the intake over that year. The compounded half-intake attrition rate from Eq. (9) can then be annualized using Eq. (4).

3.4 Intake-Free Attrition

The derivations of attrition rate formulas in the previous sections rely on the availability of intake data. However, in our original study of DSs [1] these data were not available. Only annual workforce snapshots had been available at the time, from which we could only identify *net intake* – the number of recruits, net of those that had departed before the end of the year. Access to *gross intake* was not available at that time.

Under these conditions, we measured annual attrition as the number of departing employees from the year-start workforce, divided by the year-start headcount. Here, we will call this the *intake-free* attrition rate, as does not include intake in its denominator. Let $\dot{r}$ be the net intake, then the intake-free attrition rate is

$$\dot{A}_{[0,1]} = \frac{w_0 + \dot{r} - w_1}{w_0} \tag{10}$$

which was previously called the transition wastage rate by Bartholomew *et al.* in [5].

Equation (10) represents attrition over one year. Similarly to the half-intake attrition rate, when needed over a multi-year period, the annual intake-free attrition rates were computed as

$$1 - \dot{A}_{[0,1]} = \prod_{i=1}^{n} \frac{w_i - \dot{r}_i}{w_{i-1}} \tag{11}$$

where n is the number of years, w_i is the headcount at the end of the i [th] year, and $\dot{r}_i$ the net intake over that year. The compound intake-free attrition rate from Eq. (11) can then be annualized using Eq. (4).

When intake data are available, such as in this paper, there is no reason to use the intake free attrition rate over the previously derived formulas. However, it is instructive to compare the forecasting power of such rates with that of more detailed methods described above. Indeed, we had previously relied on net intake in [1], and so we will replicate those results here for comparison.

3.5   Attrition Rates as a Function of Age

So far, we have discussed attrition rates as applicable to an entire workforce. For the purposes of this paper, we will however require attrition rates that apply to age segments within the workforce. References [10] and [11] are other works where attrition rates measured as a function of age were used in personnel modelling.

Let $[\alpha, \beta]$ be an age interval. We require an attrition rate, $A(\alpha, \beta)$, to describe departures among employees older than α and younger than β. Such a rate can be obtained by restricting attrition formula input to the particular age segment (i.e. $w_i(\alpha, \beta)$ and $r_i(\alpha, \beta)$). However, it is important to notice that intake, $r_i(\alpha, \beta)$, now encompasses more than just recruits. Indeed, if day i corresponds to an existing employee's α^{th} birthday, that employees enters the $[\alpha, \beta]$ age segment on that day, and must be counted in $r_i(\alpha, \beta)$. Similarly, on their $(\beta + 1)^{th}$ birthday, continuing employees must be subtracted from $r_i(\alpha, \beta)$ (potentially resulting in a negative overall intake count).

We estimated attrition rates using data from 1 April 2014 to 31 March 2019, for DSs 65 and older and five year ranges below that. The rates obtained from the general formula are shown in Fig. 5. Only 16 DSs were 65 and older in March 2019, which is a small segment for estimating an attrition rate. However, this segment has substantially higher attrition than any other, justifying that it be kept separate from the next-oldest segment.

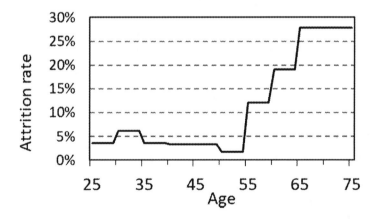

Fig. 5. Attrition rate as a function of age, obtained from the general formula.

Figure 5 follows the usual pattern seen in other DND workforces that we have studied and described as common by Bartholomew et al. [5]. Attrition is slightly higher among the youngest employees, who tend to have been recently recruited. It then stays low over several age ranges, reaching a minimum just before employees reach pension eligibility at 55. Attrition then increases steeply as increasing numbers of employees become eligible for retirement with an immediate annuity and reach the maximum number of pensionable years (35 for federal public servants).

Among public service classifications, DSs have comparatively low attrition. This is likely due to the specialized expertise of many DSs not being as readily transferrable to the wider labour market.

3.6   Attrition Rate Comparisons

The models developed in the next sections of this paper will take attrition rates as input parameters. The intake-free rate will be considered because it had been the basis for our original study [1]. The other rates are expected to yield more accurate models, as they more fully account for attrition from gross intake, but we are not aware of studies that have empirically compared the performance of these different attrition rates in modelling. Our models forward project attrition rates in a way that is based on the half-intake formula. This might justify feeding that model a rate obtained from the said formula applied to historical data. On the other hand, the half-intake formula can be interpreted as an annual approximation of the higher resolution general formula, or of the internal attrition rate. As such, it could also be reasonable to feed the model rates obtained from those formulae.

For now, Table 2 shows the attrition rates computed for the age bins considered. They are based on data from April 2014 to March 2019. We see that the general and half-intake formulas give similar rates, while the internal rate yields lower values. The intake-free rate was computed on net intake data (as opposed to the gross intake data used for the other rates), and differed most from the others.

Table 2. Comparison of the attrition rates obtained from different methods.

Age bin	General formula	Internal attrition rate	Half-intake formula	Intake-free with net intake
<30	3.6%	2.8%	3.4%	2.9%
30–34	6.2%	5.0%	6.0%	6.0%
35–39	3.7%	3.0%	3.7%	3.4%
40–44	3.3%	2.8%	3.3%	3.1%
45–49	3.4%	3.0%	3.4%	3.5%
50–54	1.7%	1.6%	1.7%	1.7%
55–59	12.0%	10.8%	12.2%	13.8%
60–64	19.1%	16.1%	19.4%	22.0%
65+	27.9%	25.0%	29.0%	41.0%

4   Workforce Age Equilibrium

In the previous sections, we have looked at the historical age distribution at hiring, and at historical attrition rates as a function of age. These historical patterns will now serve as a starting point for looking at future workforce age distributions. In this section, we will model the eventual workforce age equilibrium. In the following section, we will look at changes from year to year in workforces that are not at equilibrium.

4.1   Equilibrium Model Based on Gross Intake

Let h be the number of employees hired in a given year, and $\bar{h}(\alpha)$ the proportion of hired employees who are α years old. Then, $h \cdot \bar{h}(\alpha)$ employees of age α are hired that year. From Eq. (8), the half-intake formula for attrition, we get an expected number of departing employees aged α that year as

$$w(\alpha - 1) \cdot \bar{A}(\alpha) + h \cdot \bar{h}(\alpha) \cdot \frac{\bar{A}(\alpha)}{2} \tag{12}$$

for an expected annual attrition rate of $\bar{A}(\alpha)$ among employees aged α. We denote the attrition rate in Eq. (12) $\bar{A}$ to emphasize that it is to be an annual rate. Note also that the $w(\alpha - 1)$ employees at the start of the year will be α years old by the end of the year, and that they could be subject to a different attrition rate at each age. On average, those employees will be $\alpha - 1$ years old for half the year, and α years old for the second half. Thus, $\bar{A}(\alpha)$ should be the mean between the rates derived for employees at ages $\alpha - 1$, and α in Sect. 3.5.

Alternatively, we could have used Eq. (7), the internal attrition rate, to obtain an expected number of departing employees. However, this would have required knowing the precise timing of all intake events. Without this timing, we must make a modelling assumption about the distribution of the intake. In [6], we empirically demonstrated that the internal rate best matches the general attrition rate when we assume half of the intake occurs at the beginning of the year, and the other half at the end. When that assumption is applied to a single year, the half intake formula is obtained, which is what we used to derive Eq. (12).

At equilibrium, the workforce is neither growing nor shrinking. Thus, the number of hires must exactly match the number of departures:

$$h = \sum_{\alpha} \left[w(\alpha - 1) \cdot \bar{A}(\alpha) + h \cdot \bar{h}(\alpha) \cdot \frac{\bar{A}(\alpha)}{2} \right] \tag{13}$$

Equation (13) can now be re-worked to obtain an annual count of hires at equilibrium:

$$h = \frac{\sum_{\varphi} w[(\varphi - 1) \cdot \bar{A}(\varphi)]}{1 - \sum_{\delta} \frac{\bar{h}(\delta) \cdot \bar{A}(\delta)}{2}} \tag{14}$$

Now, in any given year, the workforce aged α will be the sum of the retained workforce that was one year younger in the previous year, and the retained recruits over the course of the previous year. Based on Eq. (8), retention among the recruits would be at a rate of $1 - \bar{A}(\alpha)/2$, yielding:

$$w_i(\alpha) = w_{i-1}(\alpha - 1) \cdot (1 - \bar{A}(\alpha)) + h \cdot \bar{h}(\alpha) \cdot \left(1 - \frac{\bar{A}(\alpha)}{2}\right) \tag{15}$$

At equilibrium, the workforce age distribution is stable from year to year. Thus, the indices in Eq. (15) can be omitted. Substituting Eq. (14) into Eq. (15) and re-arranging, we get that

$$w(\alpha) = w(\alpha - 1)(1 - \bar{A}(\alpha)) + \frac{1 - \frac{\bar{A}(\alpha)}{2}}{1 - \sum_\delta \frac{\bar{h}(\delta)\bar{A}(\delta)}{2}} \cdot \bar{h}(\alpha) \sum_\varphi w(\varphi - 1)\bar{A}(\varphi) \tag{16}$$

Given a hire age distribution ($\bar{h}(\alpha)$) and attrition rates as a function of age ($\bar{A}(\alpha)$), Eq. (16) defines a set of linear equations in the variables $\{w(\alpha)\}$: the workforce equilibrium age distribution. To solve the system, one additional constraint is required

$$\sum_\alpha w(\alpha) = w \tag{17}$$

That is, that the number of employees of each age add up to the total headcount.

When the system of linear equations defined by Eqs. (16) and (17) is solved with $\bar{h}(\alpha)$ based on five recent years (shown in Fig. 3) and $\bar{A}(\alpha)$ the general formula column in Table 2, we obtain the equilibrium distribution shown in Fig. 6.

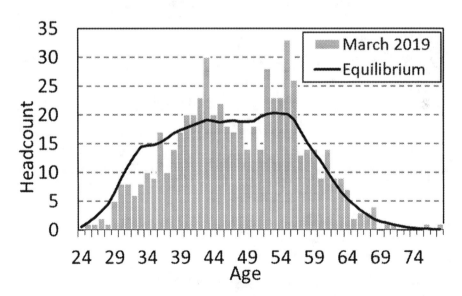

Fig. 6. Equilibrium age distribution, compared to the recent age distribution.

The solid black line in Fig. 6 shows an equilibrium age distribution that has similar, albeit smoothed, shape to the true age distribution taken at 31 March 2019. Notably, there are more employees distributed in the early thirties, and fewer in the early fifties.

This can be attributed to a recent hiring strategy focused on lower level DSs (and indirectly, younger ages). The mean age obtained from the equilibrium age distribution shown in Fig. 6 is 47.1, which is over one year younger than 48.4, the mean age of the true distribution on 31 March 2019.

Table 3. Means of the equilibrium age distributions obtained under each scenario.

Scenario	A	B	C	D	E	Current
Equilibrium mean age	42.9	44.0	45.3	45.8	46.5	47.1

Table 3 similarly shows the mean age of the equilibrium distributions obtained when $\bar{h}(\alpha)$ is taken from the scenarios described in Table 1. Each of these scenarios, which involve shifting hiring toward junior DS levels, result in younger equilibrium work-forces. Figure 7 compares the youngest of these alternative equilibrium age distributions (scenario A), to that previously shown in Fig. 6. We see how shifting hiring towards the junior levels would eventually rejuvenate the workforce.

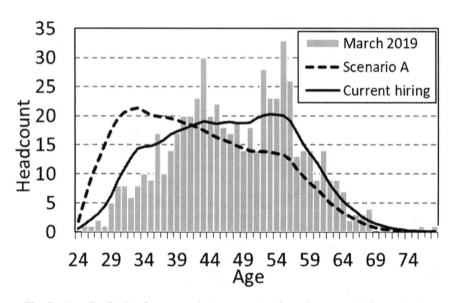

Fig. 7. Age distribution from scenario A versus that from the current hiring scenario.

4.2   Net Intake Equilibrium Model

As previously explained, our original DS study had to rely on more limited net intake data, not the gross intake used to derive the equilibrium model above. Here, we will compare our previous equilibrium model based on net intake, to see the impact of the refinement.

When working with net intake, attrition from in-year hires is not tracked, so the attrition rate only measures losses from the year-start headcount. Similar to Eq. (12), the net hiring must exactly compensate attrition at equilibrium:

$$\dot{h} = \sum_{\alpha} \left[w_i(\alpha - 1) \cdot \dot{A}(\alpha) \right] \tag{18}$$

Equation (15), which expresses the number of employees of a given age as the sum of those retained from the previous year with those hired during the year is also simplified by not having to consider attrition from the intake:

$$w_i(\alpha) = w_{i-1}(\alpha - 1) \cdot \left(1 - \dot{A}(\alpha) \right) + \dot{h} \cdot \bar{h}(\alpha) \tag{19}$$

Similarly to what was done for the gross intake model, Eqs. (18) and (19) can be combined to yield the equilibrium model that had been used in [1].

$$w(\alpha) = w(\alpha - 1)\left(1 - \dot{A}(\alpha) \right) + \bar{h}(\alpha) \sum_{\varphi} w(\varphi - 1)\dot{A}(\varphi) \tag{20}$$

It can be noted that the difference between Eqs. (16) and (20) is the added factor at the beginning fo the second term in Eq. (16). The factor is a ratio between the retention rate among in-year hires aged α, $1 - \bar{A}(\alpha)/2$, to the retention rate over all in-year hires. This factor is close to unity for most of the considered age intervals, and so it is expected that the net-intake equilibrium model will not give results that are very different from those of the gross-intake equilibrium model. Indeed, when both models are fed attrition rates and hire age data from April 2014 to March 2019, the difference between equilibrium age distributions is small (see Fig. 8). Because it is based on a more complete accounting of personnel flows, the gross intake model should be more accurate. However, given the small difference observed in Fig. 8, we expect the results from net intake to be *good enough* to inform hiring policy decisions, as was done in ref [1].

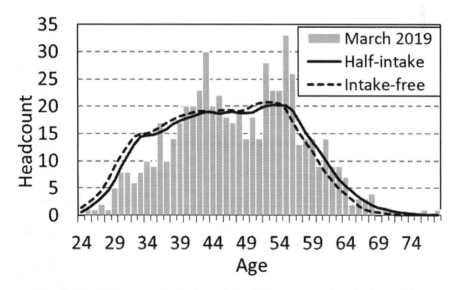

Fig. 8. Equilibrium age distributions obtained from gross and net intake models.

5   Dynamic Model

The model of the previous section allows us to forecast the eventual age distribution of a workforce equilibrium. However, it does not say how long it will take to converge, and how the age distribution will vary along the way. HR planning horizons are likely to be much shorter than the time needed for the age distribution to converge, so there is value in predicting what will happen over the shorter term.

5.1   Gross Intake Dynamic Model

Equation (16) can be adapted from an equilibrium to dynamic model. One must simply add indices to the headcounts, so they can be tracked over time:

$$
\begin{aligned}
w_i(\alpha) = & \\
& w_{i-1}(\alpha - 1)(1 - \bar{A}(\alpha)) + \frac{1 - \frac{\bar{A}(\alpha)}{2}}{1 - \sum_\delta \frac{h(\delta)\bar{A}(\delta)}{2}} \cdot \bar{h}(\alpha) \sum_\varphi w_{i-1}(\varphi - 1)\bar{A}(\varphi)
\end{aligned}
\tag{21}
$$

Feeding Eq. (21) the $\bar{h}(\alpha)$ based on five recent years (shown in Fig. 3) and the $\bar{A}(\alpha)$ from the general formula column in Table 2, we get a forecast of DS age distributions over successive years under the assumption that hire ages and attrition behaviours remain constant. A forecast of the DS mean age obtained in this way is shown in Fig. 9.

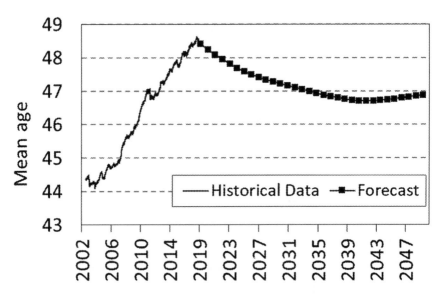

Fig. 9. Mean age forecast based on the last five years' attrition rates and hire age distribution.

The sequence of forecasted mean ages shown in Fig. 9 eventually converges to 47.1, as predicted by the equilibrium model shown in Table 3, but takes many years to get there. HR planners are likely to be more interested in the progression over the next five to ten years, as shown in Fig. 6.

5.2   Model Validation with Historical Data

It is now reasonable to ask how accurate we expect the predictions from our dynamic model to be. In our original study, sufficient data to validate the model had not been available, but we now have data that spans 16 years – enough for several years' worth of forecasting and validation. We have used successive spans of five years of historical data to fit the model, and then compare its output to the historical record. In other words, we are forecasting the past, using data from the more distant past.

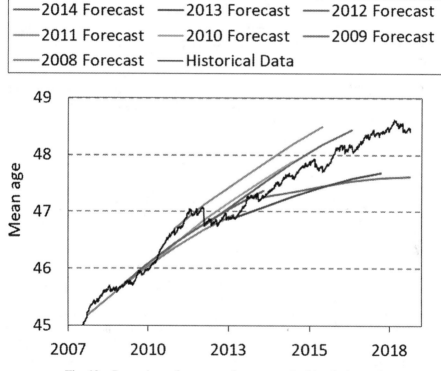

Fig. 10. Comparison of mean age forecasts to the historical record.

Figure 10 shows seven five-year forecasts of mean DS age compared to the historical values. The Forecasts are labeled by the last of the 5 fiscal years used in the generation of intake distribution and attrition rates. For example, the forecast denoted "2009 Forecast" is based on hire age and general formula attrition rates obtained from data for April 2004 to March 2009. From this, Eq. (21) was used to forecast age distributions for the next 5 years, i.e. from March 2010 to March 2014 for the "2009 Forecast". We see that the forecasts generally follow a trend that is similar to that observed in the historical record. Of course, this only works when past trends continue into the future. The most divergent forecast shown in Fig. 10 can be clearly seen as the "2014 Forecast", which doesn't capture the same upward trend characterized by the other forecasts (and historical record). This forecast was fit with data from 2009 to 2014, a period that coincided with a significant reduction in the number of DSs, and that is thus fairly different from the period of stability that followed.

It should be noted that Fig. 10 only shows mean age, which might hide some differences in the way the ages were distributed between forecasts and historical observations. We examine these differences below.

5.3   Choice of Attrition Rate Formula

So far, we have been applying the general formula to determine attrition rates for our dynamic model forecasts. However, in Sect. 3, we defined alternative attrition rates and provided some intuitive justification for their use in modelling. We will now compare prediction accuracies coming from these alternative attrition rates.

For the intake-free rate, the above dynamic model is inappropriate, so we will first review the dynamic model derived for this rate in our original work [1]. If net intake is the only data available, then Eq. (18) will govern hiring. Like with the equilibrium case, the resulting dynamic model is based on an equation similar to that for gross intake, but with the factor in front of the second term eliminated:

$$w_i(\alpha) = w_{i-1}(\alpha - 1)\big(1 - \dot{A}(\alpha)\big) + \bar{h}(\alpha) \sum_{\varphi} w_{i-1}(\varphi - 1)\dot{A}(\varphi) \tag{22}$$

We can now compare the net intake model using the intake-free attrition rate to the gross intake model using the other rates. Figure 11 shows how well forecasts perform, including those that were used in Fig. 10. The age distributions were binned in 5-year increments, as per Table 2. The sum squared differences between the age distributions of employees in each forecasted bin and the historically observed distributions was then computed as a measure of forecast accuracy. Figure 11 shows the mean sum squared difference for each forecasting method, where a different attrition rate was used. The seven 5-year time periods from Sect. 5.2 were again used here, namely; 2003–2008, 2004–2009, 2005–2010, 2006–2011, 2007–2012, 2008–2013, and 2009–2014. The results of each of the seven forecasts were compared with the historical distribution, and the average sum squared difference was obtained over all time periods to obtain a line in Fig. 11. Thus, the value shown for forecasted year '1' on the x-axis is the average of the values from the 2009 forecast obtained from 2003-08 data, 2010 forecast from 2004-09 data, and so on up to the 2015 forecast from 2009-14 data. Likewise, forecasted year '2' is the average of the second year out from the data, i.e. 2010 from the 2003-08 data, and so on. The graphs in Fig. 11 are thus indications of the accuracy of each forecast, where errors closer to 0 are more accurate.

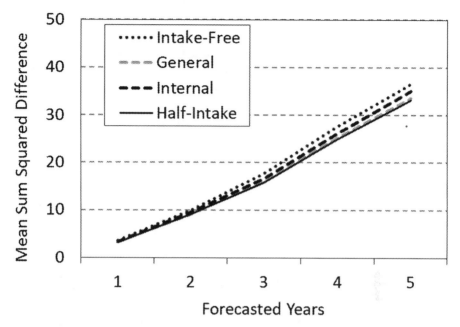

Fig. 11. Mean sum of squared differences between actual and forecasted age distributions.

As should be expected, Fig. 11 shows average sum squared differences that increase as one gets farther away from the original data used to fit the model. This can be attributed to the cumulative differences between forecast and historical data from year to year, and because hire age distributions and attrition behavior change over time, such that past data is less reflective of future patterns the further we move into the future.

Figure 11 shows that the choice of attrition rate has a noticeable impact on forecast accuracy. The best performing forecasting model used the half-intake formula as both the method for computing attrition, and projecting into the future. It is perhaps surprising that the general formula performed second-best. The general formula explicitly includes all intake and attrition events in its formulation, and therefore provides the most accurate representation of attrition behavior amongst DSs (or any workforce population). This may be a result of the limited data sets we developed our model on, and more datasets would be necessary to gather enough empirical evidence to favour one rate over another. However that will be left for future investigations.

It is clear from Fig. 11 that the intake-free attrition rate provides the least accurate forecast. As such it appears that gross intake provides a clear improvement in forecasting and should be used, wherever possible, over net intake data (which is obtained, for example, from annual snapshots of a workforce population).

However, there is much more variability in the forecast accuracy when different historical years of data were used, and the attrition rate is held constant. To illustrate this, we compare Fig. 12 to Fig. 11, which are drawn on the same scale. Figure 12 show a minimum and maximum sum of squared differences between forecast and historical record when using the half-intake attrition rate. We see that this spread in model accuracy based on the year of the forecast is much greater than the difference in accuracy from using different attrition rate formulas.

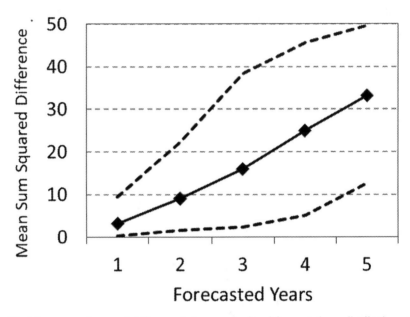

Fig. 12. Mean sum of squared difference between actual and forecasted age distributions using the half-intake model, also showing the minimum and maximum sum of squared differences.

5.4   Results

We are now ready to model how the age distribution within the DS workforce would evolve under the scenarios described in Table 1. Figure 13 shows how quickly each of the scenarios brings about a reduction in the workforce's mean age. Each of the trends converges toward the values obtained by the equilibrium model that were shown in Table 3, but some take longer than others. In [1], we showed that on the way to converging, mean age oscillates with a period of 30 to 35 years – the most common career length for DSs.

Given that it takes several years for age distributions to converge to their equilibrium following changes in hiring policy, results of the dynamic model such as those shown in Fig. 13 are helpful indicators of the near term effect of policy changes.

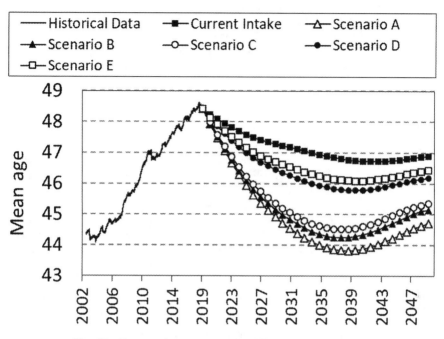

Fig. 13. Forecasted mean age under different hiring scenarios.

6 Conclusion

This paper presented approaches to measuring the effect of changes in the age distribution of hires on the age distribution of a workforce. In doing so, several conclusions can be drawn:

- Equilibrium models of a workforce's age distribution are useful to determine the direction of ongoing trends and the eventual effect of hiring policies, but it can take many years to approach this equilibrium – longer than typical HR planning horizons.
- Dynamic models of a workforce's age distribution are helpful in seeing the near term effect of changes in hiring policies.
- Access to high resolution workforce data (e.g. gross intake) resulted in improvements to our forecasts. However, further work on the robustness of this conclusion must be conducted on larger datasets.
- Among the attrition rates considered, our model accuracy was highest using the half-intake formula, followed closely by the general formula.
- The main determinant of the accuracy of our age distribution predictions appears to have been the historical years used to compute the intake distribution and attrition rate quantities, and not the method in which those quantities were calculated. It follows that forecasts will be more accurate when the historical behavior of a workforce is representative of their future behavior.

The methods presented in this paper can inform policy aimed at achieving workforce rejuvenation. Finally, in addition to workforce rejuvenation, these methods can be adapted to analyze other workforce demographic characteristics, such as gender equality.

References

1. Vincent, E.: Workforce modelling in support of rejuvenation objectives. In: Proceedings of the 8th International Conference on Operations Research and Enterprise Systems (ICORES 2019), Prague, Czech Republic, pp. 23–29 (2019)
2. Eles, P., Massel, P.: DRDC CORA's OR scientists: analysis of past hiring, career progression, and attrition trends, and development of a model to forecast future demographics, centre for operational research and analysis technical memorandum DRDC CORA TM 2006-31. Defence Research and Development Canada, Ottawa (2008). http://cradpdf.drdc.gc.ca/PDFS/unc71/p529377.pdf
3. Isbrandt, S., Zegers, A.: The arena career modelling environment individual training and education (ACME IT & E) projection tool – an overview, centre for operational research and analysis technical report DRDC CORA TR 2006-03. Defence Research and Development Canada, Ottawa (2006)
4. Erkelens, A., Isbrandt, S., Syed, F.: Development of a prototype model for civilian occupational group projections. In: Summer Computer Simulation Conference (Summer-Sim 2007), pp. 1277–1282. Curran Associates, San Diego (2007)
5. Bartholomew, D.J., Forbes, A.F., McClean, S.I.: Statistical Techniques for Manpower Planning, 2nd edn. Wiley, Chichester (1991)
6. Vincent, E., Calitoiu, D., Ueno, R.: Personnel attrition rate reporting, director general military personnel research and analysis scientific report DRDC-RDDC-2018-R238. Defence Research and Development Canada, Ottawa (2018)
7. Global Investment Performance Standard, CFA Institute (2010). https://www.gipsstandards.org/standards/Pages/currentedition.aspx
8. Okazawa, S.: Measuring attrition rates and forecasting attrition volume, centre for operational research and analysis technical memorandum DRDC CORA TM 2007-02. Defence Research and Development Canada, Ottawa (2007). http://cradpdf.drdc.gc.ca/PDFS/unc66/p527519.pdf
9. Dietz, P.: Pension Funds: Measuring Investment Performance. Columbia University, New York (1966)
10. Doumic, M., Perthame, B., Ribes, E., Salort, D., Toubiana, N.: Toward an integrated workforce planning framework using structured equations. Eur. J. Oper. Re. **262**(1), pp. 217-230 (2016). https://hal.inria.fr/hal-01343368/document
11. Foran, D., Straver, M.: Forecasting CAF releases and population: age-based attrition modelling, director general military personnel research and analysis scientific report DRDC-RDDC-2017-R176. Defence Research and Development Canada, Ottawa (2017)

A Genetic Algorithm Approach to Multi-Agent Mission Planning Problems

Branko Miloradović[(✉)], Baran Çürüklü, Mikael Ekström,
and Alessandro V. Papadopoulos[iD]

School of Innovation, Design and Engineering, Mälardalen University,
Högskoleplan 1, 721 23 Västerås, Sweden
{branko.miloradovic,baran.curuklu,mikael.ekstrom,
alessandro.papadopoulos}@mdh.se

Abstract. Multi-Agent Systems (MASs) have received great attention from scholars and engineers in different domains, including computer science and robotics. MASs try to solve complex and challenging problems (e.g., a mission) by dividing them into smaller problem instances (e.g., tasks) that are allocated to the individual autonomous entities (e.g., agents). By fulfilling their individual goals, they lead to the solution to the overall mission. A mission typically involves a large number of agents and tasks, as well as additional constraints, e.g., coming from the required equipment for completing a given task. Addressing such problem can be extremely complicated for the human operator, and several automated approaches fall short of scalability. This paper proposes a genetic algorithm for the automation of multi-agent mission planning. In particular, the contributions of this paper are threefold. First, the mission planning problem is cast into an Extended Colored Traveling Salesperson Problem (ECTSP), formulated as a mixed integer linear programming problem. Second, a precedence constraint reparation algorithm to allow the usage of common variation operators for ECTSP is developed. Finally, a new objective function minimizing the mission makespan for multi-agent mission planning problems is proposed.

Keywords: Multi-Agent Systems · Multi-agent mission planning · Extended Colored Traveling Salesperson (ECTSP) · Genetic algorithms

1 Introduction

The popularity of the multi-agent systems (MAS) keeps increasing due to advances in key technologies such as, sensors, energy sources, computing units,

This work was supported by the project Aggregate Farming in the Cloud (AFarCloud) European project, by the Swedish Foundation for Strategic Research under the project "Future factories in the cloud (FiC)" with grant number GMT14-0032, with project number 783221 (Call: H2020-ECSEL-2017-2), and by the Knowledge Foundation with the FIESTA project.

G. H. Parlier et al. (Eds.): ICORES 2019, CCIS 1162, pp. 109–134, 2020.
https://doi.org/10.1007/978-3-030-37584-3_6

and wireless communication, among others [10]. The immediate consequences of these advances are improved price-performance ratio, diverse product range, and availability of components, and finally, improved complete systems. This in turn pushes advances in application domains, which results in a pull effect, resulting in better technologies. A recent example in this regard is the Unmanned Aerial Vehicle technologies (UAV), and applications, such as inspection of critical infrastructures, monitoring construction sites and natural environments, also critical applications such as search and rescue [44]. However, the airborne problem domain is not the only one where MASs are used. Ground domain (e.g., search and rescue missions [12]) and underwater domain (e.g., seabed mapping [24]) missions benefit widely from the use of MASs.

In this context, planning consists of assigning all the tasks in a mission to agents in such a way that the plan is feasible, given a global mission objective, resources, and constraints. As in many cases, agents can have different equipment, e.g., in terms of sensors, actuators, or batteries. This results in a heterogeneous MAS configuration. Tasks in a plan can have precedence constraints (PC) in addition to equipment, or sensor requirements. The planning problem may also include choosing the optimal set of agents for the mission. Task precedence is only considered within the agent's plan, i.e., one agent cannot execute a task that precedes a task assigned to another agent. This problem is modeled as a novel TSP variation formally proposed in this paper.

This type of combinatorial problem can be solved by an exact method or with a meta-heuristic approach. Exact methods guarantee an optimal solution, usually by mapping a problem into a tree or a graph. Later, search through the nodes, prune infeasible branches and backtrack from dead-ends. Meta-heuristics solve problems differently, which sometimes leads to a sub-optimal solution. In addition, as the search space increases most planning approaches fail to produce a plan within a reasonable time. This is especially important in scenarios that require immediate access to the plan for the progress of the mission. In the general case, however, the time taken for an initial plan to be produced is less critical compared to that of re-planning. If re-planning takes too long, the state of the MAS can change while the planning process is being done. In these cases, where the necessity for a feasible plan outweighs the need for optimality, suboptimal planners are usually preferred. The planner proposed in this work is based on a Genetic Algorithm (GA) [16], and has been adapted to the specific problem of heterogeneous multi-agent mission planning, which is the focal point of this paper.

In this paper, the problem of mission planning for MAS is tackled. The main contributions of this paper are to: (i) Cast the multi-agent mission planning problem to a novel Extended Colored Traveling Salesperson Problem (ECTSP), and formulate such problem as a Mixed-Integer Linear Programming (MILP) problem; (ii) Develop a reparation algorithm to allow usage of common variation operators for ECTSP; and (iii) Propose a new optimization criterion useful for multi-agent mission planning problems. Furthermore, examples from the applications of Autonomous Underwater Vehicles (AUVs) and some randomly generated generic multi-agent missions are used in order to illustrate different contributions.

A preliminary version of this work appeared in Miloradović et al. [32]. Additional contributions are: (i) the precedence constraint representation within the MILP problem, following the paradigm of two-commodity network flow model, (ii) the description of the proposed objective function has been further detailed and analyzed, and (iii) the results section has been enriched with a more extensive evaluation of the proposed solution, also in terms of cost and required computational resources.

In Sect. 2 an overview of the related work is given. A formal problem formulation is given in Sect. 3. The genetic solver is explained in Sect. 4. Results are presented in Sect. 5, and finally Sect. 6, concludes the paper.

2 Related Work

In order to classify MAS problems, Gerkey and Matarić [13] proposed a domain independent taxonomy with three decoupled axis for multi-robot task allocation (MRTA) problems. Authors make several assumptions in the paper, however, the most important one is that tasks are assumed to be independent of each another. This means that there are no ordering constraints, synchronization or any other interrelatedness between tasks. In order to overcome this restriction, Korsah et al. [21] extended MRTA with the degree of interrelatedness. Later, Nunes et al. [38] decomposed this dimension further into synchronization and precedence constraints. The term *synchronization constraints* (SC) is used when there are temporal relations between tasks and *precedence constraints* (PC) when there are only ordering relations. PC can be seen as a subset of SC, however the focus, in this paper, will be explicitly on PC. In Operations Research these constraints are usually indicated as General Precedence Relationships [35]. The TSP variation that will be described, and addressed in this paper, can be labeled as Single-Task robots, Single-Robot tasks, Time-Extended Assignments with Synchronization and Precedence (ST-SR-TA-PC) with robots being heterogeneous.

The aforementioned task allocation is solved by using planning algorithms (planners). A planner breaks a mission plan into smaller pieces, tasks, that are sent to the appropriate agent for execution, i.e., it does the allocation of tasks to appropriate agents. Most of the approaches presented here are domain independent or can be easily translated into domain independent solutions, thus can be used in different scenarios, such as a group of Autonomous Underwater Vehicles (AUVs) or ground vehicles.

The mission planning problem for a swarm of AUVs without task interrelatedness can be solved using multi-objective harmony search algorithm [24]. A research framework on mission planning for swarms of UAVs has been proposed by Zhou et al. [48]. The problem of mission planning for a swarm of UAVs can be solved using the algorithm approach as shown in [40, 42]. The problem is modeled as a constraint satisfaction problem and solved using multi-objective GA. This work has been further extended to utilize re-planning and analysis of operator training in the control center [43]. For a similar problem of a mission planning for cooperative UAV teams, a solution was proposed by Bello-Orgaz et al. [4], that use GA with a weighted linear combination of the mission's makespan and the fuel consumption as the optimization criterion. This approach was further improved by Ramirez-Atencia et al. [41], by using weighted random generator strategies for the creation of new individuals. All of the problems in these papers can be seen as variations of the TSP.

Khamis et al. [20] presented a survey on MRTA problems, connecting the planning problem to the multiple TSP problem. In addition, the authors provide a comprehensive overview of centralized and decentralized approaches for solving MRTA. The original TSP formulation is expressed as an integer linear program and it was introduced by Dantzig and colleagues [9]. By this day many different approaches were developed and proposed in order to solve TSP. In today's state of the art two inexact approaches are on average ahead of the rest Lin-Kernighan (LKH) [19] and GA with Edge Assembly Crossover (EAX) [46]. The original problem definition is later extended to an mTSP [3]. An approach using sub-tours was proposed by Giardini et al. [14] to solve multiple TSP (mTSP). The idea was to divide a graph into subgraphs which are solved using GA. Each subgraph represents a tour for one of the salesmen. Another extension of the original TSP is done by adding Precedence Constraint (TSPPC) [22]. mTSP and TSPPC were later combined into an mTSPPC [47], although a formal problem formulation was not given.

A two-commodity flow formulation of TSP with Time Windows was given by Langevin et al. [11,25] solving problem instances of up to 60 cities. This formulation of the TSP is extended to handle precedence constraints [36] and solved by using GA. It is shown that other TSP/VRP variants can be extended in the similar manner as well [15,23]. Recently, serial [31], radial [26], and a more complete version of a Colored TSP (CTSP[1]) has been proposed by Meng et al. [30] in order to solve multiple bridge machine planning in industry.

In this paper, their approach has been further generalized by adding Two-Commodity Flow model for PC, multi depots (with different starting and ending points). This means that the solution to the problem is a Hamiltonian path and not a cycle, thus ECTSP is an open variant of TSP.

3   ECTSP Formulation

Before introducing the theoretical background of the ECTSP, the connection between the theoretical model and multi-agent mission will be explained in more detail.

3.1   Mapping Between MRTA and ECTSP

In MRTA, cities and salespersons correspond to tasks and agents, respectively. A salesperson's colors map to an equipment type of an agent, e.g., camera, gripper,

---

[1] CTSP is an abbreviation used in the literature for a Clustered TSP as well.

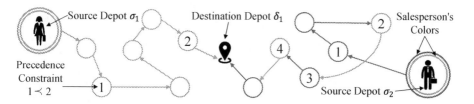

Fig. 1. An illustration of the ECTSP.

and different types of sensory equipment. Following the same rules, a color, which is associated with a task, indicates the required equipment, i.e., equipment an agent should have in order to successfully complete that task. In contrast to the classical TSP formulation, tasks are neither instantaneous nor have a predefined duration. The duration of a task is estimated based on the agent's capabilities. Moreover, equipment heterogeneity is not the sole determinant for the differences between the agents. Every agent may have a different speed, therefore the duration of task execution may depend on the selected agent. In addition to equipment requirements and duration, tasks may have an additional parameter, i.e., precedence constraints. Some tasks might need to be completed before or after some other task in a mission. Precedence constraints in ECTSP are essentially an ordering constraint. This means that interrelated tasks will be allocated to the same agent. Following Korsah's definition [21], it can be concluded that ECTSP has only intra-schedule dependencies. Although actions are durative, this problem in its core is temporally simple [8], i.e., although actions have a duration, actions are ordered, and no concurrent actions are possible.

For example, Fig. 1 shows two salespersons starting from two different source depots (σ_1 and σ_2, respectively). Each of them has two different colors and visits a certain number of tasks with matching colors. Every salesperson ends its tour in the destination depot δ_1. Cities that have precedence constraints are marked with numbers in the figure (1 has to be visited before 2, i.e., $1 \prec 2$). The need for precedence constraint is apparent, as explained in this multi-agent mission example: Let's assume that an agent has two tasks, e.g., survey an area with camera for data collection, and send data back. Obviously, these two tasks are interrelated—the data cannot be sent before it is acquired. Thus the only possible ordering between the two tasks is to gather the data first, before sending it to the receiving destination. Note, further, that there is no constraint on the delay between the end of the first task, and the start of the second task.

3.2   Problem Formulation

Suppose that an ECTSP has m salespersons, $s \in \mathcal{S} := \{s_1, s_2, \ldots, s_m\}$, n cities, $v \in \mathcal{V} := \{v_1, v_2, \ldots, v_n\}$, k colors, $c \in \mathcal{C} := \{c_1, c_2, \ldots, c_k\}$, q source depots $\sigma \in \Sigma := \{\sigma_1, \ldots, \sigma_q\}$, and w destination depots, $\delta \in \Delta := \{\delta_1, \ldots, \delta_w\}$ where $m, n, k, q, w \in \mathbb{N}^+$. Each salesperson s starts from a source depot σ and finishes its tour at a destination depot δ. Source and destination depots are not considered to be cities. The superset containing all of the cities $\mathcal{V}$ and depots is defined as $\widetilde{\mathcal{V}} := \mathcal{V} \cup \{\Sigma, \Delta\}$. In addition, for the simplicity, a superset containing all source depot and city elements is defined as $\mathcal{V}^\Sigma := \mathcal{V} \cup \Sigma$. In the same manner a superset containing all elements of destination depot and cities is defined as $\mathcal{V}^\Delta := \mathcal{V} \cup \Delta$. This problem can be formulated over a directed graph $\mathcal{G} = (\widetilde{\mathcal{V}}, \mathcal{E})$, where $\mathcal{E} : \widetilde{\mathcal{V}} \times \widetilde{\mathcal{V}} \mapsto \mathbb{R}_0^+$. An edge $e \in \mathcal{E}$, connecting vertexes $i, j \in \widetilde{\mathcal{V}}$ can be expressed as

$$e(i, j) = \begin{cases} w_{ij}, & \text{if } i \text{ is connected to } j \\ 0, & \text{otherwise,} \end{cases}$$

where $w_{ij} \geq 0$ represents the cost of edge $e(i, j)$. The decision variable $x_{ijs} \in \{0, 1\}$ can be defined as

$$x_{ijs} = \begin{cases} 1, & \text{if } s \in \mathcal{S} \text{ travels from } i \in \widetilde{\mathcal{V}} \text{ to } j \in \widetilde{\mathcal{V}}, \\ 0, & \text{otherwise.} \end{cases}$$

Every city $i \in \mathcal{V}^\Sigma$ has a weight $\xi(i)$, $\xi : \mathcal{V}^\Sigma \mapsto \mathbb{R}_0^+$ (with $\xi(i) = 0$ when $i \in \Sigma$). Also, every city $i \in \mathcal{V}$ is associated with a color $f_c(i)$, with $f_c : \mathcal{V} \mapsto \mathcal{C}$. Each salesperson $s \in \mathcal{S}$ has a set of colors $\mathcal{C}_s \subseteq \mathcal{C}$ assigned to it—source and destination depots do not have colors. In contrast to city color matrix that was defined by Li et al. [26], here a color matrix of a salesperson s, $\mathcal{A}_s \in \{0, 1\}^{n \times n}$, shows openness of cities towards a salesperson s, and is defined as $\mathcal{A}_s := [a_{ijs}]$, with

$$a_{ijs} = \begin{cases} 1, & f_c(v_i) \in \mathcal{C}_s \wedge f_c(v_j) \in \mathcal{C}_s \wedge \pi_{ij} = 1 \\ 0, & \text{otherwise,} \end{cases}$$

where $\Pi = [\pi_{ij}]_{n \times n}$ is the adjacency matrix indicating the precedence relations among the cities, where $\pi_{ij} = 1 \iff i \prec j$, and 0 otherwise. The definition of the color matrix $\mathcal{A}_s$ can be extended to include the depots as:

$$\bar{a}_{ijs} = \begin{cases} a_{ijs}, & i, j \in \mathcal{V}, \\ 1, & (i \in \Sigma, j \in \mathcal{V}^\Delta) \vee (i \in \mathcal{V}^\Sigma, j \in \Delta), \\ 0, & (i, j \in \Sigma) \vee (i, j \in \Delta). \end{cases}$$

A salesperson s is allowed to only visit the cities specified in its extended color matrix $\mathcal{A}_s$:

$$x_{ijs} \leq \bar{a}_{ijs}, \qquad \forall i \in \mathcal{V}^\Sigma, \forall j \in \mathcal{V}^\Delta, \forall s \in \mathcal{S}, i \neq j. \tag{1}$$

Furthermore, only one salesperson $s \in \mathcal{S}$ can enter (Eq. 2) and leave (Eq. 3) each:

$$\sum_{s \in \mathcal{S}} \sum_{i \in \mathcal{V}^\Sigma} x_{ijs} = 1, \qquad \forall j \in \mathcal{V}, i \neq j, \tag{2}$$

$$\sum_{s \in \mathcal{S}} \sum_{j \in \mathcal{V}^\Delta} x_{ijs} = 1, \qquad \forall i \in \mathcal{V}, i \neq j. \tag{3}$$

The final destination of a salesperson s is always a destination depot:

$$\sum_{i \in \mathcal{V}^{\Sigma}} \sum_{j \in \Delta} x_{ijs} = 1, \qquad\qquad \forall s \in \mathcal{S}. \qquad (4)$$

Note that some salespersons can go directly from a source depot to a destination depot, i.e., $x_{ijs} = 1, i \in \Sigma, j \in \Delta$. This means that the salesperson is not used in the final plan.

The starting location of a salesperson s is always a source depot:

$$\sum_{i \in \Sigma} \sum_{j \in \mathcal{V}^{\Delta}} x_{ijs} = 1, \qquad\qquad \forall s \in \mathcal{S}. \qquad (5)$$

The number of salespersons $\mathcal{B}_{\sigma}$ in each source depot $\sigma \in \Sigma$ is given, and it is such that $\sum_{i \in \Sigma} \mathcal{B}_i = |\mathcal{S}|$, and:

$$\sum_{s \in \mathcal{S}} \sum_{j \in \mathcal{V}^{\Delta}} x_{ijs} = \mathcal{B}_i, \qquad\qquad \forall i \in \Sigma, \qquad (6)$$

And it defines the initial deployment of the salespersons over the source depots. In order to ensure that the same agent enters and exits a certain city:

$$\sum_{i \in \mathcal{V}^{\Sigma}} x_{ijs} = \sum_{k \in \mathcal{V}^{\Delta}} x_{jks}, \qquad\qquad \forall j \in V, \forall s \in S. \qquad (7)$$

A salesperson s cannot travel from a city i to the same city i:

$$x_{iis} = 0, \qquad\qquad \forall i \in \tilde{\mathcal{V}}, \forall s \in \mathcal{S}. \qquad (8)$$

The description above concludes the formulation of the Colored TSP with multiple source and destination depots, and heterogeneous salespersons. The extension with intra-scheduled dependencies is presented below by using Two-commodity Network Flow (TNF) paradigm.

3.3   Two-Commodity Network Flow Model for Precedence Constraints

As a simple example, assume that salesperson s delivers a full container of some commodity (e.g., a bottle of gas), and picks up an empty container, for each customer. At all times, the salesperson will have a total of n containers (full + empty). A salesperson s has two distinct commodities $y_{ijs}, z_{ijs} \in \mathbb{N}^0$, when moving from city i to city j, in the network with n_s number of nodes (cities), with

$$n_s = \sum_{i \in \mathcal{V}^{\Sigma}} \sum_{j \in V} x_{ijs}, \qquad\qquad \forall s \in \mathcal{S}. \qquad (9)$$

The commodity y_{ijs} is supplied by n_s units at a selected starting node, $i \in \Sigma$, and consumed by one unit at each node that is not the starting node. On the other hand, the commodity z_{ijs} is consumed by n_s units at the starting node, and supplied by one unit at the other nodes. This kind of network flow of commodities is described by two properties:

(i) Sum of all commodities y_{ijs} and z_{ijs} in any feasible tour must be equal to n_s

(ii) The quantity of commodity y_{ijs} exiting from a node is decreasing by one unit, while the quantity of commodity z_{ijs} is increasing by one unit as the tour proceeds.

The flow conservation equations for the commodity y_{ijs} and z_{ijs}, are defined as follows:

$$\sum_{j \in \mathcal{V}^\Delta} y_{ijs} - \sum_{j \in \mathcal{V}^\Sigma} y_{jis} = - \sum_{j \in \mathcal{V}^\Delta} x_{ijs}, \qquad \forall i \in \mathcal{V}, \forall s \in \mathcal{S}. \qquad (10)$$

$$\sum_{j \in \mathcal{V}^\Delta} z_{ijs} - \sum_{j \in \mathcal{V}^\Sigma} z_{jis} = \sum_{j \in \mathcal{V}^\Delta} x_{ijs}, \qquad \forall i \in \mathcal{V}, \forall s \in \mathcal{S}. \qquad (11)$$

In addition to the flow conservation equations (Eqs. 10 and 11), it is necessary to define the flow conservation equations for the edges going from the set Σ of source depots.

$$\sum_{i \in \Sigma} \sum_{j \in \mathcal{V}} y_{ijs} - \sum_{i \in \mathcal{V}} \sum_{j \in \mathcal{V}} y_{ijs} = \sum_{i \in \mathcal{V}^\Sigma} \sum_{j \in \mathcal{V}} x_{ijs}, \qquad \forall s \in \mathcal{S}, \qquad (12)$$

$$\sum_{i \in \Sigma} \sum_{j \in \mathcal{V}} z_{ijs} - \sum_{i \in \mathcal{V}} \sum_{j \in \mathcal{V}} z_{ijs} = - \sum_{i \in \mathcal{V}^\Sigma} \sum_{j \in \mathcal{V}} x_{ijs}, \qquad \forall s \in \mathcal{S}. \qquad (13)$$

Finally, if node i must precede node j precedence constraints can be imposed as follows:

$$\sum_{k \in \mathcal{V}} y_{iks} \geq \sum_{l \in \mathcal{V}} y_{jls}, \qquad \forall i, j \in \mathcal{V}, \forall s \in \mathcal{S} : i \prec j, \qquad (14)$$

$$\sum_{k \in \mathcal{V}} z_{iks} \leq \sum_{l \in \mathcal{V}^\Delta} z_{jls}, \qquad \forall i, j \in \mathcal{V}, \forall s \in \mathcal{S} : i \prec j. \qquad (15)$$

Finally, we introduce big-M constraints:

$$y_{ijs} \leq M \cdot x_{ijs}, \qquad \forall i \in \mathcal{V}^\Sigma, \forall j \in \mathcal{V}^\Delta, \forall s \in \mathcal{S}, \qquad (16)$$

$$z_{ijs} \leq M \cdot x_{ijs}, \qquad \forall i \in \mathcal{V}^\Sigma, \forall j \in \mathcal{V}^\Delta, \forall s \in \mathcal{S}. \qquad (17)$$

Let us assume that node i precedes node j ($i \prec j$), then three different precedence relationship cases between nodes can be defined:

1. Node j must be visited exactly β nodes after node i,
2. Node j must be visited in less than β nodes after node i,
3. Node j must be visited in more than β nodes after node i.

For the simplicity, and without loss of generality, the focus will be only on the case 3, with β equal to 1, described in (Eqs. 14 and 15). This implies that city j can be visited any time, as long as it is after city i.

If there is a constraint that implies that node j must be visited immediately after node i, then this is the case 1 where β is equal to 1.

In this case, edge contraction can be used to reduce the size of a graph. For a graph $G = (V, E)$ and an edge $e = (i, j) \in E$, $i \neq j$, the contraction of e in G is an operation that results in the graph $G' = (V', E')$, where the resulting graph G' has one less edge than G. If S defines a set of edges that are incident to vertices i and j, which are replaced with a single vertex v, then $S' = \{S \setminus e\}$ is a set of edges incident to v. After edge contraction, the weight of the new vertex is a sum of weights i and j, added to a weight associated with the cost ω_{ij} of going from i to j, i.e., $\xi_x = \xi_i + \xi_j + \omega_{ij}$. The same approach can be used in the case of n consecutive nodes with PC.

Few assumptions are made in this model. Tasks with ordering constraints have the same color, i.e., the same salesperson can perform these tasks. In addition, tasks can only have one PC relation. Ordering constraints cannot be shared between the salesmen, i.e., if node i precedes node j, both nodes are visited by the same salesperson s. Omniscience is assumed, as well as atomic tasks, and deterministic effects.

3.4   Objective Function

MAS is a structure composed of multiple interacting agents, in which the agents fulfill various goals, or have specific purposes. In the context of this work, there is an overall goal that will be achieved by an agent population when they perform the tasks that are assigned to them. This overall goal is mathematically represented with a function commonly known as the objective function, that must be either maximized or minimized. An overview of the most common objective functions in MRTA problems is given by Nunes *et al.* [38]. As can be seen in this work, a mission can have many different objective functions, however, the most common optimization criterion is the minimization with respect to time, i.e., the total mission duration. Other common optimization criteria are the minimization of the path, or energy consumption.

The duration of a mission is a straightforward concept in terms of single-agent missions. However, a mission duration can be defined in many different ways in the context of the MAS. A mission management tool is used for defining missions to be planned with the planner presented in Sect. 4. Figure 2 shows the obtained plans for different objective functions described below.

One definition is the sum of all tasks in a mission (miniSum) [28]:

$$f_s = \min_x \sum_{s \in \mathcal{S}} \sum_{i \in \mathcal{V}^{\Sigma}} \sum_{j \in \mathcal{V}^{\Delta}} (\omega_{ij} + \xi(i)) x_{ijs}. \tag{18}$$

The interval from the start of the first task to the end of the last task (miniMax) is another common definition [4]:

$$f_m = \min_x \max_s \sum_{i \in \mathcal{V}^{\Sigma}} \sum_{j \in \mathcal{V}^{\Delta}} (\omega_{ij} + \xi(i)) x_{ijs}. \tag{19}$$

Nonetheless, neither of these approaches is suitable for the problem presented in this paper. The former approach can produce a plan where one agent is doing much more work than the other agents (see Fig. 2a), where only one agent (indicated in red) is used and it covers all tasks. In an actual underwater mission involving AUVs, this can mean that the extraction vessels and the crew have to stay in the open sea for a longer time, thus increasing the overall cost of the

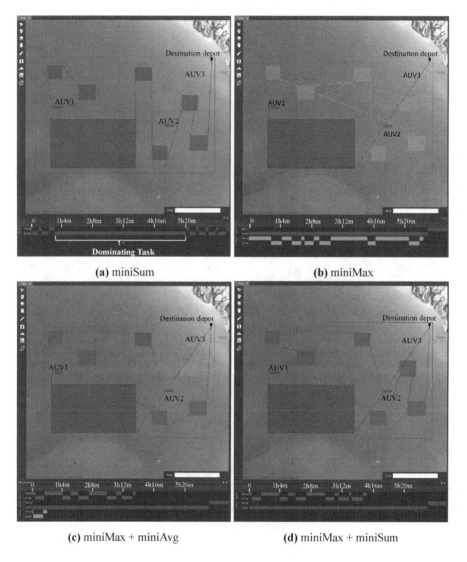

(a) miniSum   (b) miniMax

(c) miniMax + miniAvg   (d) miniMax + miniSum

Fig. 2. Different planning outputs based on four different objective functions. Identical sets of AUVs are assumed in all four cases. Bottom part of the figures shows the complete plans represented as Gantt charts.

mission. The latter approach minimizes the maximum cost of an agent over all the agents. This approach works well if tasks are instantaneous or have the same duration. However, if there is a task that dominates all the other tasks, e.g., its duration is longer than the makespan of any other agent's plan, then the mission will not be optimized at all, as shown in Fig. 2b. It can be observed that AUV3's (yellow coloured agent) makespan is shorter than the duration of the one single task belonging to AUV2 (red coloured agent) (Fig. 2b, see the bottom part of the figure showing the Gantt chart representing the plan). In this case, the planner would try to minimize the longest plan belonging to the agents, and since that plan would consist of one very long task it wouldn't be possible for the planner to minimize the complete plan further. In addition, the other agents that have shorter makespans would just be ignored. This issue is partly mitigated in [1] where a linear aggregation function is used and the optimization criterion was to minimize the maximum and average task completion times (miniMax + miniAvg), and total idle times. This approach favors the usage of as many agents as possible, although it may mean increase in cost for every agent added to the solution. Such a solution is shown in Fig. 2c. Thus, the final plan includes all three available agents.

All these objective functions have their usefulness in different cases, or scenarios. However, as discussed above, they also have weaknesses. In order to address these problems, in this paper a new optimization criterion has been proposed (Eq. 20). It is defined as a weighted combination of miniSum (Eq. 18) and miniMax (Eq. 19):

$$J = w_1 f_m + w_2 f_s, \tag{20}$$

subject to constraints from (1) to (17), where $w_1, w_2 > 0$ are user defined weights. How this objective function affects the final solution is shown in Fig. 2d.

4   Genetic Mission Planner (GMP)

Multi-agent missions can be conducted in a harsh and challenging environment (e.g., underwater, or airborne, missions) where many different problems may arise. Since some of these problems cannot be foreseen beforehand, changes in the mission, while it is being executed, are common. Although the initial plan making is not bounded by time, since no agent needs to be deployed, the re-planning is. Re-planning, in this context, can be seen as planning again with new initial conditions. Since multi-agent missions are usually costly and autonomy of agents is limited (especially in airborne missions), the re-planning process should be very fast. Since this is an NP-hard problem, exact solution algorithms do not scale well. This is one of the reasons for the use of a meta-heuristic approach over exact methods. GA and its numerous variations have been widely used for solving combinatorial optimization problems such as TSP [27,37,45], Vehicle Routing Problem (VRP) [2,29], job shop scheduling problems [5,17], and resource constrained problems [6,18].

In this paper, the encoding of chromosomes is done in a same manner as in Miloradović *et al.* [34]. Two arrays of integer identifiers are used to represent different types of genes. Task genes and agent genes are encoded in the first array, whereas the second array consists of task parameter genes (Fig. 3). Tasks have 4 parameters: (i) precedence relations, (ii) required equipment, (iii) task duration, and (iv) task location. The length of a chromosome can be fixed prior to the mission if the length of a plan is known, i.e., the number of tasks and the number of agents to be used is fixed. However, if a planner can introduce new tasks or decides not to use all available agents, variable chromosome length is preferred (Brie *et al.* [6]). In this paper, a mix of these two approaches is used, i.e., chromosome length is fixed, however, besides Active Genes (AG), i.e., tasks and agents, a certain number of inactive dummy genes is introduced. The reason for this is that, although the number of tasks in a mission is fixed, the planner is allowed to optimize the number of agents involved. This implies that the number of dummy genes is in the range from 0 to $(m - 1)$, while the chromosome length is fixed to the size of $(n + m)$.

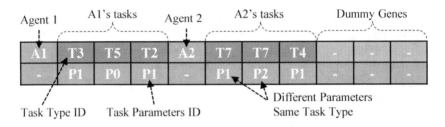

Fig. 3. An example of an individual in the population.

The initial population is generated with the respect to given constraints, however, the task allocation process is randomized. At the beginning of the search process, the minimum number of agents necessary for the successful completion of the mission is determined. Then a number of agents in the chromosome is randomly picked in the range of that determined minimum and the maximum number of available agents. Assume a multi-UAV mission during which patches of agricultural fields will be scanned through the execution of 10 tasks. Assume further that some of these tasks require a thermal camera, whereas others humidity sensors. In addition, there are three agents available, one with a thermal camera and the other two with humidity sensors. Then the number of agents per chromosome in the initial population will range from 2 to 3. After this step, a list of possible tasks for every agent is created. The number of tasks per agent is randomly determined based on the previously created list. Two or more agents may have the ability to do the same task, e.g., a task requires a thermal camera, and there are three agents with thermal camera available. In this specific case, a task is randomly allocated to one of the valid agents. Task allocation is repeated until there are no more tasks left to allocate and the whole process is repeated for every chromosome in the population.

4.1 Variation Operators

The two main variation operators of a GA are crossover and mutation. Various different types have been proposed for solving GA TSP [7] and VRP problems [39].

Crossover. In a classical implementation of a GA for TSP, chromosomes are arrays consisting of genes (city identifiers). Crossover operators can mix individuals in many possible ways and produce feasible solutions. The only restriction is city duplicates in a potential solution.

N-point crossover operators usually produce offspring which violate given constraints, thus a repair procedure is usually required. This led to the development of an ordered based (OB) crossover operators, which tend to retain ordering from parents in the process of creating offspring like Partially Mapped Crossover (PMX). As a result, sub-tours are passed to the offspring without change. This type of operator creates a non-repeating sequence of cities thus, there is no need for a repairing procedure. Another crossover operator that shows promising results when applied to the TSP problem is Edge Recombination Crossover (ERX). The idea is to preserve edges between nodes from parents and pass them to the offspring. However, for the ECTSP, the above-mentioned procedure is less trivial, since precedence and color constraints need to be taken into consideration, in order to avoid the creation of infeasible solutions.

Mutation. The mutation is the source of variability as it allows genetic diversity in the population. Every individual has a certain probability to be selected for mutation. In this work, two types of mutation schemes are introduced. One operates on the task genes, whereas the other mutates agent genes.

Task gene mutation consists of swap gene and inserts gene mutation operations (Algorithm 1). Task swap mutation swaps two task genes in a chromosome, meaning that it can both swap tasks within a single agent or between two agents. In the first case, only the ordering of tasks changes. In the second case, tasks are exchanged between the two agents. This is done with respect to color constraints, i.e., tasks can be exchanged only between agents capable of performing those tasks. Insert mutation randomly chooses a task gene, removes it from its current location and inserts it in a new location in a chromosome. Similarly to the task gene mutation, the insertion can be within the same agent or to a different one. While performing insert mutation, color constraints are respected as in the previous case.

Agent genes mutate in two different ways, by adding agent genes to the chromosome (growth mutation) or by removing agent genes from the chromosome (shrink mutation) (Algorithm 2). Agent growth mutation adds a new agent gene to the plan if the limit of available agents is not reached in that specific chromosome. The new agent gene is inserted at a random location in the chromosome. After insertion of a new agent gene, all tasks from that location to the next agent gene, or end of the chromosome are allocated to the newly added agent. If the

---

Algorithm 1. Task Swap/Insert Mutation.

1: **procedure** TASKMUTATION(*population*)
2:     Select random chromosome i from the pop.
3:     Select random task gene from i
4:     **case (1)** - Task Swapping
5:       Create a list of valid tasks for swapping
6:       Choose a task gene randomly from that list
7:       Swap tasks
8:     **case (2)** - Task Insertion
9:       Create a list of suitable agents for chosen task
10:      Randomly select agent and position and insert gene from step 3.
11: **return** modified chromosome

---

new agent may not be able to perform any of its tasks, randomly reallocation to other agents is performed, keeping the feasibility of the plan intact. Agent shrink mutation removes one agent from a chromosome, reallocating its tasks to other agents if possible, i.e., if such action do not break the feasibility of the plan. If such a scenario cannot be avoided the agent shrink mutation is skipped for that instance. Removing the only agent from a plan or removing the only agent with the required equipment for a specific task are examples in this regard.

Algorithm 2. Agent Growth/Shrink Mutation.

1: **procedure** AGENTMUTATION(*population*)
2:     Select random chromosome i from the pop.
3:     **case (1)** - Agent Growth
4:       **If** there are available agents
5:         Select random position to insert new agent
6:         Validate that no constraint is violated
7:         Add new agent
8:         Check if new agent's plan is valid
9:         If not, reassign invalid tasks to other agents
10:    **case (2)** - Agent Shrink
11:      **If** the minimum is not reached
12:        Select random agent gene for removal
13:        Remove that gene
14:        Reassign its tasks to other agents with the respect to constraints
15: **return** modified chromosome

Both gene and agent mutation algorithms consider color constraints ensuring that the mutation process does not produce infeasible solutions with respect to colors. However, neither the crossover nor the mutation operators have the ability to produce feasible solutions with respect to precedence constraints. In order to overcome this issue, the Precedence Constraint Reparation (PCR) algorithm was developed.

4.2   Precedence Reparation Algorithm

The PCR is applied once after the creation of the initial population. Then it is applied in every generation after crossover (when used) and mutation. The algorithm (Algorithm 3) starts by iterating through allocated tasks for each agent. If a conflict is identified, the algorithm continues searching through tasks until the type of conflict is recognized. Two conflict types can occur. Firstly, when both tasks that have precedence relations are allocated to the same agent, and at the same time the ordering is wrong. Secondly, when tasks are not allocated to the same agent. This type has three sub-cases, which are explained in more detail below.

In case (1) task genes are simply swapped. In case (2) there are three different sub-cases. In sub-case (1) both tasks are allocated to agents able to handle defined constraints, then a uniform random mechanism is used to determine which of the tasks will be re-allocated and which one will retain the previous allocation. In sub-case (2) one of the tasks is allocated to an agent that cannot fulfill the task's constraints. In that case, the invalid task is allocated to the agent of the task with valid constraints. In sub-case (3) both of the tasks are allocated to agents that cannot fulfill their constraints. In this case, a search is performed to find if there is an agent available that can fulfill both of the tasks. If that agent exists, both tasks are re-allocated to that agent. In every case, the location of the re-allocated task within an agent is randomly determined with respect to the PC. The graphical representation of how PCR works is given in Fig. 4. An example of the aforementioned cases with possible solutions will be described in detail in the next paragraph.

Algorithm 3. Precedence Constraint Reparation.

1: **procedure** PCR(*population*)
2:      Iterate through the chromosome and find a task with a PC violation
3:      **case (1)** - Tasks are allocated to the same agent
4:        Swap conflicting tasks
5:      **case (2)** - Tasks are not allocated to the same agent
6:        *sub-case (1)* - Both agents are suitable for the allocated tasks
7:            Randomly choose which task to re-allocate
8:            Move chosen task with respect to the PC
9:        *sub-case (2)* - One agent is not suitable for the allocated task
10:           Reallocate invalid task to the agent of a valid task w.r.t. PC
11:       *sub-case (3)* - Both agents are unsuitable for the allocated tasks
12:           Create a list of suitable agents
13:           Randomly select an agent
14:           Reallocate both tasks to that agent w.r.t. PC
15: **return** *fixed population*

Assume that task T5 has to be executed before task T8, i.e., T5 $\prec$ T8, and that only agents A1 and A2 have the necessary equipment to perform these

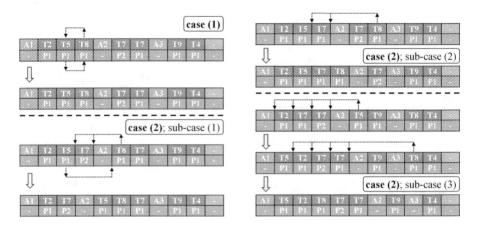

Fig. 4. An example of the mechanism of PCR, showing both cases and all three possible sub-cases of case (2).

tasks. In case (1) both tasks are allocated to the appropriate agent (A1), and a simple task swap is performed in order to fix the precedence conflict. The first sub-case of the case (2) shows the problem when each task is allocated to a different agent, however, both A1 and A2 are able to execute the allocated task. Now the option is to either move T8 to A1 after T5 or T7, or as an alternative, move T5 to A2 and place before T8 (the latter option is shown in Fig. 4). In the sub-case (2) T8 is allocated to A3 although A3 does not meet the requirements of T8. The solution includes moving T8 to A1, and place either after T5 or T7 (the latter option is shown in Fig. 4). Sub-case (3) shows a problem when both of the tasks are allocated to agents that do not meet the tasks' requirements. In this case, first the task that has precedence (T5) over the other task (T8) is reallocated to an appropriate agent, in this case A1. In the example in Fig. 4, T5 is placed right after agent gene A1. After this is done T8 has to be reallocated to A1, as well. Possible locations of insertion of T8 are after T5, T2, T7, or another T7 task. In Fig. 4, T8 is shown to be inserted after T2. As already mentioned, when making a decision of where to re-allocate a certain task uniform random function is used.

4.3   Fitness Function

The fitness function evaluates every individual in the population by calculating a quantitative measure of its performance referred to as fitness, or cost. The actual search is done in a solution space consisting of either only feasible, or both feasible and infeasible solutions. Allowing infeasible solutions in the population is a design choice.

The first approach is used in this paper with respect to mutation operator. This operator is defined in such a way that only feasible solutions can be produced. However, crossover operator may produce infeasible solutions since it does not take into account color or precedence of tasks. In this case, where operators are not bound to always produce valid chromosomes, the fitness function has a penalty/award system in order to deal (penalize) with invalid chromosomes [33]. The former approach performs a search over reduced solution space, whereas in the latter approach implementation of variation operators is simpler.

In the fitness function, the candidate solution that is being evaluated is first divided into agent's plans, i.e., sub-plans per each agent. Each plan is evaluated separately and results are combined afterward to form an overall chromosome cost. The objective function that is being optimized is given in Eq. (20).

5   Simulation Results

The experimental platform was i7-8700 @ 4.4 GHz with 16 GB of DDR4 RAM. The GMP is implemented in the C++ programming language as single-threaded process. Ten different and increasingly difficult multi-agent mission scenarios[2] are randomly generated and used for evaluating the proposed approach. Tasks can have three different colors. Agents are randomly created with up to three different colors. Instances range from 10 to 500 tasks and 1 to 10 agents. The amount of precedence constraints per instance is randomly generated in a range from 5–20% of total task number. The objective function (Eq. 20) is used with w_1 and w_2 being 1 and 0.1, respectively. Every instance is run 30 times. The detailed instance settings are given in Table 1.

Table 1. Settings for different simulation scenarios.

Instance	Task's settings						Agent's settings					
	Eq1	Eq2	Eq3	#PC	#Tasks	ATD	Eq1	Eq2	Eq3	#Agents	SD	DD
1	–	10	–	1	10	594.7	–	1	–	1	1	1
2	–	10	20	5	30	552.5	–	1	2	2	2	1
3	27	23	–	5	50	485.5	2	2	–	3	3	2
4	21	23	31	13	75	491.6	1	3	3	4	4	2
5	36	33	31	6	100	528.9	4	2	2	5	5	3
6	44	52	54	25	150	519.9	3	3	5	6	6	3
7	58	65	77	14	200	475	5	2	1	7	7	4
8	89	102	109	51	300	518.4	2	5	5	8	8	4
9	149	127	124	60	400	515.8	7	3	5	9	9	5
10	172	163	165	30	500	500.4	5	6	7	10	10	5

*(#PC - Number of precedence constraints; #Tasks - The total number of tasks; ATD - Average Task Duration in seconds; #Agents - The total number of available agents; SD - The number of source depots; DD - The number of destination depots)

Due to space limitations, only selected representative results will be shown. This includes the comparison between different mutation probabilities. Among tested values (5–30%), 10% is the lowest mutation probability that does not lead

---
[2] The link to benchmark scenarios: https://github.com/mdh-planner/ECTSP.

to the degradation of performance. For this reason, in all the following benchmarks the mutation probability will be fixed to 10%. In the same way, it is found that the lowest percentage for the individuals to be kept and transferred to the next generation (elitism) is 5%. Three different crossover operators were tested: (i) single point, (ii) partially mapped, and (iii) edge recombination crossover. While single point crossover performed poorly, as expected on combinatorial problems, edge recombination crossover beat partially mapped crossover by a small margin. This is an expected result, since ERX tries to maintain edges connecting two tasks. This comes particularly handy in the case of precedence constraints. In the following tests, ERX was used, with different probability factors.

To determine the best combination of variation operators, four different combinations were tested (Table 2).

It can be seen that the gap between settings "X, M, & PC" and the others increase as the complexity of the instances increase. For simpler instances, the difference in best and median values is not large, except for the setting with crossover and mutation without precedence reparation algorithm. This setting performed really poorly on all instances but the first one, and that is the reason it will be excluded from further discussion. The difference between the best values for the other three settings for the first 5 instances ranges from 0% to 7%. This gap increases for harder instances, especially in the case of only mutation. The solutions this setting produces for harder instances lags behind the other two settings. From this test it can be concluded that only "M & PC" and "X, M & PC" settings are competitive. The difference between these two settings never exceeded 10%. This required a closer look so a new set of tests was conducted to determine the best setting for provided benchmark.

The new test consisted of comparisons between a setting with "M & PC", "X(10%), M, & PC", and "X(70%), M, & PC". This time, the population size has been increased to 500. Obtained results are presented in Table 3.

Table 2. Simulation results for 10k generation and 200 population over different variation operators settings.

Inst.	Different variation operators settings															
	M [$\times 10^5$]				M & PC [$\times 10^5$]				X(70%) & M [$\times 10^5$]				X(70%), M, & PC [$\times 10^5$]			
	best	std.	mdn.	t [s]	best	std.	mdn.	t [s]	best	std.	mdn.	t [s]	best	std.	mdn.	t [s]
1	0.79	0	0.79	2.60	0.79	0	0.79	2.76	0.79	0.03	0.82	28.0	0.79	0	0.79	26.34
2	1.00	0.06	1.09	6.08	1.01	0.06	1.08	6.62	1.68	0.11	2.18	65.69	0.98	0.04	1.01	75.09
3	0.93	0.11	1.09	9.21	0.96	0.06	1.03	9.93	2.38	0.06	2.23	111.4	0.93	0.05	1.04	127.9
4	1.45	0.16	1.70	12.68	1.41	0.07	1.52	13.42	3.62	0.03	3.75	167.9	1.36	0.03	1.40	197.0
5	1.18	0.08	1.31	16.49	1.14	0.08	1.26	17.14	3.40	0.06	3.56	231.5	1.14	0.05	1.20	261.6
6	1.99	0.19	2.26	23.49	1.42	0.07	1.56	23.17	4.60	0.10	4.77	333.6	1.41	0.07	1.52	421.1
7	3.06	0.15	3.34	22.42	2.88	0.14	3.16	24.30	12.1	0.03	12.2	443.7	2.70	0.11	2.84	602.7
8	4.05	0.40	4.85	40.35	2.64	0.10	2.78	41.86	8.97	0.10	9.20	707.7	2.48	0.11	2.78	642.4
9	5.56	0.45	6.52	56.48	3.02	0.08	3.22	58.43	10.3	0.13	10.6	1025	2.90	0.10	3.13	1820
10	5.70	0.35	6.50	72.08	3.46	0.09	3.66	75.11	12.2	0.15	12.6	1396	3.40	0.11	3.67	1352

*(Inst. - test instance; M - Mutation; X - Crossover, PC - Precedence Constraint Reparation; mdn. - median; std. - standard deviation; best - best solution found)

The difference between results is quite small up to Instance 7, after this, the gap started increasing in the favor of "X(70%), M, & PC", while setting "X(10%), M, & PC" started performing worse than the other two. On hardest instances, the gap between "M & PC" and "X(70%, M, & PC" is less than 10%, however the difference in execution time is 16.3x faster in favor of "M & PC". In order to gain a more solid statistical insight into the difference between these three settings, a series of statistical tests were conducted.

We formulate the **null hypothesis:** *"Using different variation operator settings has no significant effect on the final result"* is true. It is assumed that the samples used in these experiments are random and independent.

Since the sample data can be highly skewed and have extended tails, an average of that data can produce a value that behaves non-intuitively, thus the median was used instead of the mean. The non-parametric Mann-Whitney-Wilcoxon test is used. Since multiple comparisons are performed, the Benjamini-Hochberg (B-H) procedure is applied in order to adjust the false discovery rate. The B-H critical value is calculated as $(i/m) \cdot Q$, where i is the individual p-value rank, m is the number of tests and Q is the false discovery rate percentage set to 1%. The variation operator settings that had the best median value is compared to all other setups in all 3 scenarios. Since there is a statistically significant difference between the results (Table 4), the null hypothesis is rejected.

Statistical tests show that for larger instances there is a benefit of using ERX crossover operator with 70% probability. In comparison with GA setting with 10% ERX probability, there is no significant statistical difference for the first 6 instances. Instances 7–10 are performed better with 70% probability of crossover. The comparison with "M & PC" is somewhat similar, except that for instance 4 there is statistically significant difference and for instance 8 there is not.

The comparison of "X(10%), M, & PC" and "M, & PC" is shown in Table 5. For instances 1, 3, and 6 there is no statistically significant difference. Instances 2, 4, 5, 7, and 8 show difference in favor of "X(10%), M, & PC", while instances 9 and 10 (marked with "*") show that there is statistically significant difference in favor of "M, & PC". This means that performance of these two settings are dependent on the instance itself and no general assumption can be made. On the contrary, setting "X(70%), M, & PC" shows solid improvement in performance over other settings as the size of the problem increases. It can be concluded that "X(70%), M, & PC" performs equally good or outperforms all other GA settings presented in this paper. The drawback is the execution time, as it gets up to 16x slower than "M, & PC" and up to 2.8x slower than "X(10%), M, & PC". If time is critical, it might be better idea to go for "M, & PC" even though the solution is a bit worse. Obvious example can be a re-planning of an ongoing mission. For the initial planning "X(70%), M, & PC" can be used to get the best possible plan, but if there is a need for a re-plan, "M & PC" can be used to get good-enough plan in a shorter period of time. These differences are graphically presented in Fig. 5, where Fig. 5a, c and e show the maximum running time for different GA settings including variation in population size and number of generations for the largest instance (Instance 10). On the other hand,

Table 3. Simulation results for 10k generation and 500 population over different variation operators settings.

Inst.	Different variation operators settings											
	M & PC [×10⁵]				X(10%) & M, & PC [×10⁵]				X(70%), M, & PC [×10⁵]			
	best	std.	mdn.	t [s]	best	std.	mdn.	t [s]	best	std.	mdn.	t [s]
1	0.79	0	0.79	8.47	0.79	0	0.79	36.10	0.79	0	0.79	29.02
2	0.98	0.06	1.08	18.82	0.98	0.03	1.03	81.69	0.98	0.03	1.01	82.09
3	0.93	0.07	1.01	27.66	0.92	0.05	1.00	128.6	0.92	0.05	1.01	138.3
4	1.38	0.07	1.48	36.01	1.36	0.02	1.40	179.0	1.36	0.02	1.40	221.8
5	1.15	0.06	1.28	45.48	1.13	0.05	1.21	120.0	1.14	0.04	1.20	319.0
6	1.38	0.06	1.52	60.77	1.45	0.04	1.54	149.0	1.38	0.06	1.51	556.1
7	2.83	0.14	3.11	65.10	2.71	0.10	2.87	181.8	2.65	0.07	2.79	837.1
8	2.49	0.09	2.68	110.1	2.59	0.09	2.74	349.3	2.46	0.08	2.61	1479
9	2.88	0.07	3.00	151.3	2.97	0.10	3.17	893.0	2.78	0.07	2.91	2341
10	3.25	0.08	3.43	194.9	3.49	0.08	3.68	1122	3.08	0.09	3.34	3172

*(Inst. - test instance; M - Mutation; X - Crossover, PC - Precedence Constraint Reparation; mdn. - median; std. - standard deviation; best - best solution found)

Table 4. Statistical test of the gathered results shown in Table 2.

X(70%), M, & PC vs.	Inst. 1 p-val.	Inst. 2 p-val.	Inst. 3 p-val.	Inst. 4 p-val.	Inst. 5 p-val.	Inst. 6 p-val.	Inst. 7 p-val.	Inst. 8 p-val.	Inst. 9 p-val.	Inst. 10 p-val.	Crit. value
X(10%), M, & PC	1	0.068	0.085	0.395	0.371	0.006	6×10^{-6}	10^{-5}	10^{-10}	4×10^{-11}	3×10^{-4}
M & PC	1	0.001	0.559	10^{-4}	0.002	0.050	8×10^{-11}	0.008	4×10^{-6}	10^{-4}	7×10^{-4}

Fig. 5b, d and f show how the median cost changes with the change of population size and number of generations for every instance for three different GA settings described in Table 3.

Finally, the convergence of the GA for "X(70%), M, & PC" setting is shown in Fig. 6. The horizontal axis shows the number of function evaluations, i.e., how many times the fitness of an individual was evaluated. The vertical axis shows cumulative probability, ranging from 0 to 1. All instances are presented, except instance 1. Instance 1 appears to be too simple, and it is solved to optimality on every run.

The Empirical Cumulative Distribution Function (ECDF) shows the convergence rate of every instance to the best known solution for that instance. Each sub-figure shows three curves with the distribution of the solutions within predefined distances from the best one: The black dotted curve shows the distribution of solutions within 5% of the best solution, the blue dashed curve shows the distribution within 10%, and finally, the red dash-dotted curve shows the distribution of solutions within 15% from the best known solution.

By analyzing the ECDF, a few things can be noticed. First, it gives deeper insight into the distribution of the solutions. Secondly, it provides insights on how to tune the algorithm parameters. If a given range from the best solution is reached late in the optimization process (e.g., Instance 9 and 10) it means that the algorithm would probably benefit if the number of generations is increased.

Table 5. Statistical test with p-value = 0.05.

X(10%), M, & PC vs.	Inst. 1 p-val.	Inst. 2 p-val.	Inst. 3 p-val.	Inst. 4 p-val.	Inst. 5 p-val.	Inst. 6 p-val.	Inst. 7 p-val.	Inst. 8 p-val.	Inst. 9 p-val	Inst. 10 p-val.
M & PC	1	0.021	0.530	$2 \cdot 10^{-4}$	0.012	0.706	$5 \cdot 10^{-8}$	0.027	$5 \cdot 10^{-8}*$	$10^{-10}*$

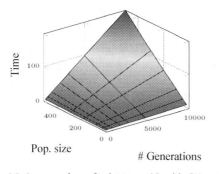

(a) Average times for instance 10 with GA set to 10% mutation and no crossover.

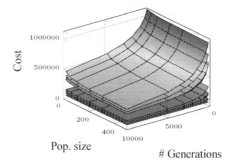

(b) Median cost for all instances with GA set to 10% mutation and no crossover.

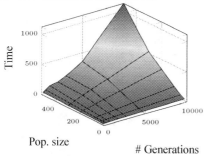

(c) Average times for instance 10 with GA set to 10% mutation and 10% crossover.

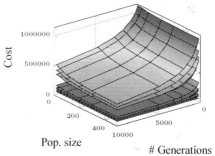

(d) Median cost for all instances with GA set to 10% mutation and 10% crossover.

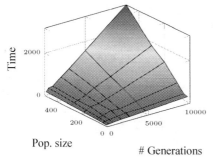

(e) Average times for instance 10 with GA set to 10% mutation and 70% crossover.

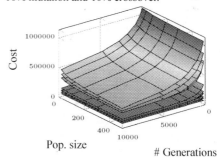

(f) Median cost for all instances with GA set to 10% mutation and 70% crossover.

Fig. 5. Analysis of computation time and cost for different GA settings.

On the other hand, if the value of 1, for the cumulative probability, is reached very fast, the number of generations can be decreased. The analysis of the ECDF can also help in understanding the underlying behavior of the algorithm on different instances, and give more insight on what are the "hard" instances for a specific algorithm. For example, Instance 3 is very hard to solve, since only less than 10% of the solutions are within 5% distance from the best one. While Instance 4, which has more tasks, reaches distance of 5% in 90% of the times. This kind of analysis can be also misleading in a sense that maybe the best known solution for Instance 3 is in fact optimal solution, while the best known solution for instance 4 could be a sub-optimal solution found in local minimum, where most of the population converged. This analysis also shows that the hardness of the instance does not necessarily increase with the increase of number of tasks and agents.

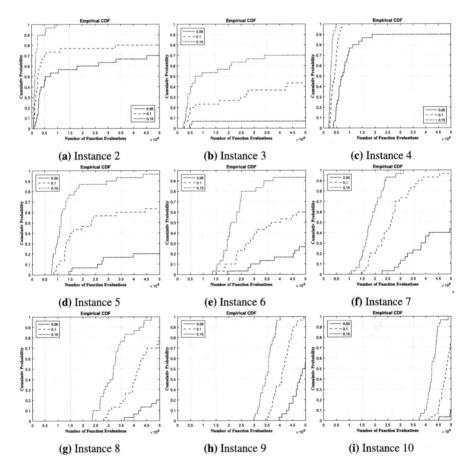

Fig. 6. Empirical CDF for instances 2–10 with GA settings: 10% mutation, 70% crossover, 500 population size, and 10000 generations.

6   Conclusion

This paper proposed a genetic algorithm for the automation of multi-agent mission planning. First, it was shown how multi-agent mission planning can be cast into a MILP formulation. The novel formulation is based on the colored TSP and it is named ECTSP. It is concluded that ECTSP can be used to model different multi-agent missions, from different real-world domains. In this work, a GA was designed to be used for ECTSP's objective function optimization with improvements in the variation operators aimed at handling given constraints. As demonstrated, the objective function presented in this work is more suitable in specific multi-agent mission cases than other solutions found in the literature. In addition, a PCR algorithm is presented that helps improve the convergence of the GA.

By analysis of the presented benchmark results for different GA settings it can be concluded that the best results are achieved when both mutation and crossover operator (ERX) are used combined with PCR algorithm. In addition, if there is a need for a good tradeoff between convergence time and quality of the solution, the GA can be run with only mutation and PCR. Benchmark results show that the speed gain can be up to 16x, with the degradation in the cost of up to 10%.

Acknowledgements. Special thanks to Afshin E. Ameri for developing GUI for the MMT.

References

1. Alighanbari, M., Kuwata, Y., How, J.P.: Coordination and control of multiple UAVs with timing constraints and loitering. In: Proceedings of the 2003 American Control Conference, vol. 6, pp. 5311–5316, June 2003
2. Baker, B.M., Ayechew, M.: A genetic algorithm for the vehicle routing problem. Comput. Oper. Res. **30**(5), 787–800 (2003)
3. Bektas, T.: The multiple traveling salesman problem: an overview of formulations and solution procedures. Omega **34**(3), 209–219 (2006)
4. Bello-Orgaz, G., Ramirez-Atencia, C., Fradera-Gil, J., Camacho, D.: GAMPP: genetic algorithm for UAV mission planning problems. In: Novais, P., Camacho, D., Analide, C., El Fallah Seghrouchni, A., Badica, C. (eds.) Intelligent Distributed Computing IX. SCI, vol. 616, pp. 167–176. Springer, Cham (2016). https://doi.org/10.1007/978-3-319-25017-5_16
5. Bhatt, N., Chauhan, N.R.: Genetic algorithm applications on job shop scheduling problem: a review. In: International Conference on Soft Computing Techniques and Implementations (ICSCTI), pp. 7–14 (2015)
6. Brie, A.H., Morignot, P.: Genetic planning using variable length chromosomes. In: International Conference on Automated Planning and Scheduling, pp. 320–329. ICAPS (2005)
7. Contreras-Bolton, C., Parada, V.: Automatic combination of operators in a genetic algorithm to solve the traveling salesman problem. PLoS One **10**(9), e0137724 (2015)

8. Cushing, W., Kambhampati, S., Mausam, Weld, D.S.: When is temporal planning really temporal? In: Proceedings of the 20th International Joint Conference on Artifical Intelligence, IJCAI 2007, pp. 1852–1859 (2007)
9. Dantzig, G., Fulkerson, R., Johnson, S.: Solution of a large-scale traveling-salesman problem. J. Oper. Res. Soc. Am. **2**(4), 393–410 (1954)
10. Dorri, A., Kanhere, S.S., Jurdak, R.: Multi-agent systems: a survey. IEEE Access **6**, 28573–28593 (2018)
11. Finke, G., Claus, A., Gunn, E.: A two-commodity network flow approach to the traveling salesman problem. Congressus Numerantium **41**(1), 167–178 (1984)
12. Frasheri, M., Cürüklü, B., Ekström, M., Papadopoulos, A.V.: Adaptive autonomy in a search and rescue scenario. In: Proceedings of the 12th IEEE International Conference on Self-Adaptive and Self-Organizing Systems (SASO), pp. 150–155, September 2018
13. Gerkey, B.P., Matarić, M.J.: A formal analysis and taxonomy of task allocation in multi-robot systems. Int. J. Robot. Res. **23**(9), 939–954 (2004)
14. Giardini, G., Kalmár-Nagy, T.: Genetic algorithm for combinatorial path planning: the subtour problem. Math. Probl. Eng. 1–31 (2011)
15. Gouveia, L., Pesneau, P., Ruthmair, M., Santos, D.: Combining and projecting flow models for the (precedence constrained) asymmetric traveling salesman problem. Networks **71**(4), 451–465 (2018)
16. Holland, J.H.: Genetic algorithms. Sci. Am. **267**(1), 66–73 (1992)
17. Qing-dao-er ji, R., Wang, Y.: A new hybrid genetic algorithm for job shop scheduling problem. Comput. Oper. Res. **39**(10), 2291–2299 (2012)
18. Kadri, R.L., Boctor, F.F.: An efficient genetic algorithm to solve the resource-constrained project scheduling problem with transfer times: the single mode case. Eur. J. Oper. Res. **265**(2), 454–462 (2018)
19. Helsgaun, K.: An effective implementation of K-opt moves for the Lin-Kernighan TSP heuristic. Math. Program. Comput. **1**, 119–163 (2009)
20. Khamis, A., Hussein, A., Elmogy, A.: Multi-robot task allocation: a review of the state-of-the-art. In: Koubâa, A., Martínez-de Dios, J.R. (eds.) Cooperative Robots and Sensor Networks 2015. SCI, vol. 604, pp. 31–51. Springer, Cham (2015). https://doi.org/10.1007/978-3-319-18299-5_2
21. Korsah, G.A., Stentz, A., Dias, M.B.: A comprehensive taxonomy for multi-robot task allocation. Int. J. Robot. Res. **32**(12), 1495–1512 (2013)
22. Kubo, M., Kasugai, H.: The precedence constrained traveling salesman problem. J. Oper. Res. Soc. Jpn. **34**(2), 152–172 (1991)
23. Kuschel, T.: Two-commodity network flow formulations for vehicle routing problems with simultaneous pickup & delivery and a many-to-many structure. Inform. Kommunikationssysteme Supply Chain Manag. Logist. Transp. **5**, 153–169 (2008)
24. Landa-Torres, I., Manjarres, D., Bilbao, S., Del Ser, J.: Underwater robot task planning using multi-objective meta-heuristics. Sensors **17**(4:762), 1–15 (2017)
25. Langevin, A., Desrochers, M., Desrosiers, J., Gélinas, S., Soumis, F.: A two-commodity flow formulation for the traveling salesman and the makespan problems with time windows. Networks **23**(7), 631–640 (1993)
26. Li, J., Zhou, M., Sun, Q., Dai, X., Yu, X.: Colored traveling salesman problem. IEEE Trans. Cybern. **45**(11), 2390–2401 (2015)
27. Ma, F., Li, H.: An algorithm in solving the TSP based on the improved genetic algorithm. In: 2009 First International Conference on Information Science and Engineering (ICISE 2009), pp. 106–108, December 2009

28. Maoudj, A., Bouzouia, B., Hentout, A., Toumi, R.: Multi-agent approach for task allocation and scheduling in cooperative heterogeneous multi-robot team: Simulation results. In: 13th International Conference on Industrial Informatics (INDIN), pp. 179–184, July 2015
29. Masum, A.K.M., Shahjalal, M., Faruque, F., Sarker, I.H.: Solving the vehicle routing problem using genetic algorithm. Int. J. Adv. Comput. Sci. Appl. **2**(7), 126–131 (2011)
30. Meng, X., Li, J., Dai, X., Dou, J.: Variable neighborhood search for a colored traveling salesman problem. IEEE Trans. Intell. Transp. Syst. **19**(4), 1018–1026 (2018)
31. Meng, X., Li, J., Zhou, M., Dai, X., Dou, J.: Population-based incremental learning algorithm for a serial colored traveling salesman problem. IEEE Trans. Syst. Man Cybern.: Syst. **48**(2), 277–288 (2018)
32. Miloradović., B., Çürüklü., B., Ekström., M., Papadopoulos., A.V.: Extended colored traveling salesperson for modeling multi-agent mission planning problems. In: Proceedings of the 8th International Conference on Operations Research and Enterprise Systems - Volume 1: ICORES, INSTICC, pp. 237–244. SciTePress (2019)
33. Miloradović, B., Çürüklü, B., Ekström, M.: A genetic planner for mission planning of cooperative agents in an underwater environment. In: 2016 IEEE Symposium Series on Computational Intelligence (SSCI), pp. 1–8, December 2016
34. Miloradović, B., Çürüklü, B., Ekström, M.: A genetic mission planner for solving temporal multi-agent problems with concurrent tasks. In: Tan, Y., Takagi, H., Shi, Y., Niu, B. (eds.) ICSI 2017. LNCS, vol. 10386, pp. 481–493. Springer, Cham (2017). https://doi.org/10.1007/978-3-319-61833-3_51
35. Monma, C.: Sequencing with general precedence constraints. Discrete Appl. Math. **3**(2), 137–150 (1981)
36. Moon, C., Kim, J., Choi, G., Seo, Y.: An efficient genetic algorithm for the traveling salesman problem with precedence constraints. Eur. J. Oper. Res. **140**(3), 606–617 (2002)
37. Nian, L., Jinhua, Z.: Hybrid genetic algorithm for TSP. In: Seventh International Conference on Computational Intelligence and Security, pp. 71–75, December 2011
38. Nunes, E., Manner, M., Mitiche, H., Gini, M.: A taxonomy for task allocation problems with temporal and ordering constraints. Robot. Autonom. Syst. **90**, 55–70 (2017)
39. Puljić, K., Manger, R.: Comparison of eight evolutionary crossover operators for the vehicle routing problem. Math. Commun. **18**(2), 359–375 (2013)
40. Ramirez-Atencia, C., Bello-Orgaz, G., R-Moreno, M.D., Camacho, D.: Solving complex multi-UAV mission planning problems using multi-objective genetic algorithms. Soft Comput. **21**(17), 4883–4900 (2017)
41. Ramirez-Atencia, C., Del Ser, J., Camacho, D.: Weighted strategies to guide a multi-objective evolutionary algorithm for multi-UAV mission planning. Swarm Evol. Comput. **44**, 480–495 (2019)
42. Ramirez-Atencia, C., R-Moreno, M.D., Camacho, D.: Handling swarm of UAVs based on evolutionary multi-objective optimization. Prog. Artif. Intell. **6**(3), 263–274 (2017)
43. Ramirez-Atencia, C., Rodríguez-Fernández, V., Gonzalez-Pardo, A., Camacho, D.: New artificial intelligence approaches for future UAV ground control stations. In: 2017 IEEE Congress on Evolutionary Computation (CEC), pp. 2775–2782 (2017)
44. Valavanis, K.P., Vachtsevanos, G.J.: Handbook of Unmanned Aerial Vehicles. Springer, Heidelberg (2014)

45. Yu, Y., Chen, Y., Li, T.: A new design of genetic algorithm for solving TSP. In: Fourth International Joint Conference on Computational Sciences and Optimization, pp. 309–313, April 2011

46. Nagata, Y., Kobayashi, S.: A powerful genetic algorithm using edge assembly crossover for the traveling salesman problem. Inform. J. Comput. **25**, 346–363 (2013)

47. Zhong, W.: Multiple traveling salesman problem with precedence constraints based on modified dynamic tabu artificial bee colony algorithm. J. Inform. Comput. Sci. **11**(4), 1225–1232 (2014)

48. Zhou, X., Wang, W., Wang, T., Li, X., Li, Z.: A research framework on mission planning of the UAV swarm. In: 12th System of Systems Engineering Conference, pp. 1–6, June 2017

Combining Machine Learning and Operations Research Methods to Advance the Project Management Practice

Nikos Kanakaris, Nikos Karacapilidis, Georgios Kournetas,
and Alexis Lazanas[(✉)]

Industrial Management and Information Systems Lab, MEAD,
University of Patras, 26504 Rio Patras, Greece
{nkanakaris, kournetag}@upnet.gr,
{karacap, alexlas}@upatras.gr

Abstract. Project Management is a complex practice that is associated with a series of challenges such as handling of conflicts and dependencies in resource allocation, fine tuning of projects to avoid fragmented planning, handling of potential opportunities or threats during the execution of a project, and alignment between projects and business objectives. Traditionally, methods and tools to address these issues are based on analytical approaches developed in the realm of the Operations Research discipline. Aiming to facilitate and augment the quality of the Project Management practice, this paper proposes a hybrid approach that builds on the synergy between contemporary Machine Learning and Operations Research techniques. Based on past data, Machine Learning techniques can predict undesired situations, provide timely warnings and recommend preventive actions regarding problematic resource loads or deviations from business priority lists. The applicability of our approach is demonstrated through two real examples elaborating two different datasets. In these examples, we comment on the proper orchestration of the associated Operations Research and Machine Learning algorithms, paying equal attention to both optimization and big data manipulation issues.

Keywords: Project Management · Machine Learning · Operations Research · Intelligent optimization

1 Introduction

Project Management (PM) is a complex practice that is highly fluid and hard to predict, thus imposing a series of challenges to organizations and experts [20]. Such challenges may concern alignment between projects and their business objectives, handling of conflicts and dependencies in resource allocation, fine tuning of multiple projects to avoid fragmented planning, as well as informed and diffused decision making to handle potential opportunities or threats during the execution of a project [39].

At the same time, PM is inherently collaborative and knowledge-intensive. Issues to be addressed are characterized by ever-increasing amounts of different types of data and knowledge, which may be obtained from various sources and vary in terms of

© Springer Nature Switzerland AG 2020
G. H. Parlier et al. (Eds.): ICORES 2019, CCIS 1162, pp. 135–155, 2020.
https://doi.org/10.1007/978-3-030-37584-3_7

subjectivity, ranging from individual opinions and estimations to broadly accepted practices and indisputable measurements and results [22]. Furthermore, their types can be of diverse level as far as human understanding and machine interpretation are concerned.

Up to now, the majority of methods and tools aiming to facilitate and augment the quality of PM are based on the application of advanced analytical approaches developed and elaborated in the realm of the Operations Research (OR) discipline. These approaches employ techniques such as mathematical optimization and statistical analysis to look for optimal or suboptimal solutions to diverse PM issues. In addition, the application of Artificial Intelligence (AI) techniques to automate project management has been first proposed more than 30 years ago. At that time, the proposed AI-leveraged project management systems used knowledge processing and procedural techniques to provide new kinds of decision support for project objective-setting and control [23].

Nowadays though, the adoption of AI in the data-intensive and cognitively-complex PM settings enables a series of advancements. AI - and in particular Machine Learning (ML) - techniques can aid project managers easily delegate thousands of tasks, while sustaining a holistic view of their resources and projects. This contributes to the achievement of the required accuracy and precision when dealing with bottle-necks or constraints that may obstruct business processes. At the same time, these techniques can aid managers and experts to interpret big volumes of data and gain valuable insights towards improving their overall PM practice. Based on past data, they can predict undesired situations, provide timely warnings and recommend preventive actions regarding problematic resource loads or deviations from business priority lists [20].

Admittedly, each of the abovementioned disciplines (OR and AI) has significantly contributed to the improvement of the PM practice, by addressing the associated issues from a different philosophy and research perspective. Moreover, both disciplines elaborate a mixture of problem modeling and problem solving methods. According to Radin [30], an OR analyst must trade off tractability (i.e. "the degree to which the model admits convenient analysis") and validity (i.e. "the degree to which inferences drawn from the model hold for real systems"). At a high level, the OR and ML analysts face the same validity and tractability dilemmas and it is not surprising that both can exploit the same optimization toolbox.

However, we argue that the joint consideration of these two disciplines has not been thoroughly explored yet, and has much potential to further augment PM-related business intelligence. Such an approach will concentrate on both planning and execution of individual projects, as well as on their association with past data and their impact on the wider business. Moreover, this approach can appropriately represent and process the associated data and knowledge, while at the same time remedy the underlying cognitive overload issues. Particular attention should be also given to the expression and maintenance of tacit knowledge (i.e. knowledge that employees do not know they possess or knowledge that they cannot express with the means provided), which predominantly exists and dynamically evolves in PM settings [20].

In line with the above remarks, this paper expands on [20] by attempting to shape a hybrid approach for better handling PM issues by meaningfully integrating tools originally developed in the context of OR and AI. More specifically, in Example 1 (Sect. 4.1): (i) we describe in depth the Data attributes concerning Construction Projects as well as the involved constructors, (ii) we apply data pre-processing methods in order to retain data that contribute in gaining statically significant information, (iii) we perform the Kruskal-Wallis test to determine whether a statistically significant difference between constructors and projects' *Delay* attribute exists and (iv) we formulate the scoring function in order to provide more accurate results. Additionally, in Example 2 (Sect. 4.2): (i) we redefine the overall approach by using classification instead of clustering methods and we provide more in depth analysis of each step, (ii) we increase the success chances of the project by properly assigning the available developers to each project issue, (iii) we examine our approach using real data contrary to hypothetical data used in [20] and (iv) we evaluate our approach by utilizing the *Local Surrogate Models (LIME)* explanation method [15, 32] in order to get a solid understanding of the underlying mechanism of our trained model.

The remainder of the paper is organized as follows: Sect. 2 discusses background work considered in the context of our approach, which is analytically described in Sect. 3; the applicability of the proposed approach is demonstrated through two realistic examples in Sect. 4; the contribution of our approach is discussed in Sect. 5; finally, concluding remarks and future work directions are outlined in Sect. 6.

2 Background Issues

Numerous software solutions to project management exist in the market nowadays. The list of the most widely adopted ones includes Wrike (www.wrike.com), Asana (www.asana.com), Trello (www.trello.com), and Jira (www.atlassian.com/software/jira). These solutions offer a user-friendly environment that mainly enables issue tracking and supports various PM functions. In addition, by providing interactive graphics, issue boards and timelines, they simplify planning, collaboration, reporting and time management. It is broadly admitted that existing commercial PM solutions may increase an organization's productivity and prevent the teams from diverging from their actual goals. However, they unintentionally hide important PM-related information, due to the complex multidimensional data found in the hosted projects.

At the same time, by adopting an AI-perspective, a range of digital project management assistants has been already developed, including solutions such as Stratejos.ai (www.stratejos.ai), PMOtto.ai (www.pmotto.ai), and x.ai (www.x.ai). This category of solutions is based on seamless, easy-to-use interfaces that assist project managers in common tasks (e.g. a project's supervision). They rely on the expressiveness, immediacy, interactivity and descriptiveness that natural language provides to offer a 'zero-level' entrance environment. They are used to automate repetitive work such as creating project's tasks by analyzing textual conversations, to remind and organize important events such as meetings, to extract shallow insights (e.g. 'top contributors of the week'), and to answer simple queries (e.g. 'what is my team working on today?').

We argue that this second category of solutions offers narrow predictions and automations. In particular, their underlying reasoning mechanisms mainly build on rules to store and manipulate knowledge, and ignore contemporary AI technologies that

can uncover insights, perform more complex tasks, make explainable recommendations, and support informed decision making, sometimes in ways that outperforms what
people are able to do today. Furthermore, each of these digital personal assistants is
relevant to a specific project management need (e.g. reporting, scheduling meetings,
organizing events); thus, they are unable to embrace a 'single-access-point' approach
that mitigates the overall PM complexity.

From an OR perspective, a series of techniques and tools have been proposed and
extensively used to solve various PM related issues. OR techniques provide solutions
in problems such as prediction, resource allocation, forecasting, scheduling, task
assignment, networking etc. These techniques are supported by very useful software libraries such as *pyschedule* (github.com/timnon/pyschedule), *PuLP*
(github.com/coin-or/pulp), Google OR-tools (developers.google.
com/optimization), JuMP.jl [12], Hungarian.jl (github.com/Gnimuc/
Hungarian.jl) and CVXPY (www.cvxpy.org). These libraries support a variety
of OR techniques including integer, linear, convex and dynamic programming. However, these techniques tend to add more complexity on the overall PM practice, mainly
due to the complicated mathematical models needed to operate. Another drawback is
that these techniques are unable to learn by the system's experience, which often results
to the proposition of optimal or near-optimal solutions that are not realistically feasible.

With the advent of big data and cloud computing era, machine learning techniques
gain ground in a variety of scientific and commercial sectors. These techniques (and
corresponding algorithms) can categorize items, predict values, identify meaningful
relationships, and detect data patterns or unexpected behavior (anomaly detection). ML
approaches are usually grouped into four categories, namely supervised learning, semi-
supervised learning, unsupervised learning and reinforcement learning [14].

Supervised learning refers to the process of learning aiming to predict values (e.g.
house prices) or classify items into categories (e.g. categories of projects) by using
labelled training data. Common algorithms and methods used in supervised learning
include k-nearest neighbors, naive Bayes, decision trees, linear regression, and support
vector machines. Semi-supervised learning combines both labeled and unlabeled input
data for training, where in most cases there is a small amount of labeled data and a huge
amount of unlabeled data available. Unsupervised learning analyzes unlabeled data to
identify patterns or cluster similar items into groups using alternative distance metrics
(e.g. Euclidean distance, Manhattan distance). Common algorithms used in unsupervised learning include k-means, DBSCAN, OPTICS, Apriori and hierarchical clustering [3, 40]. Finally, reinforcement learning approaches iteratively interact with their
environment to identify specific actions that maximize a reward or minimize a risk.
Common algorithms and methods used in this category include Q-learning, temporal
difference, and deep adversarial networks. The above ML techniques and algorithms
are fully supported today by various software libraries and environments, such as *scikit-
learn* [28], *H2O.ai* [8], *Tensorflow* [1], *PyTorch* [27] and *WEKA* [17].

As a last note, it is worth mentioning that most AI-based approaches to PM build on
artificial neural networks. Related works discuss how neural networks are capable to
assist project managers in problems such as resource allocation, prediction, clustering,
classification [7] and forecasting [44]. Neural network techniques have been also
applied to predict construction cost and schedule success [43]. Other representative

works concern development of a neural network to estimate project performance [9], or to classify the level of a project's riskiness by exploiting the knowledge extracted from data concerning past successful and unsuccessful projects [10]. An interesting overview of the different types of neural network models applied in business can be found in [37].

3   The Proposed Approach

In ML, generalization is the most essential property used to validate a novel approach. For a practical ML problem, the analyst might pick one or more families of learning models and an appropriate training loss/regularization function, and then search for an appropriate model that performs well, according to some estimate of the generalization error based on the given training data [11]. This search typically involves some combination of data preprocessing, optimization and heuristics [34]. Every stage of the process can introduce errors that can degrade the quality of the resulting inductive functions. In the related literature, particular attention is paid to three sources of such errors. The first source of error is due to the fact that the underlying true function and error distribution are unknown, thus any choice of data representation, model family and loss functions may not be suitable for the problem and thus introduce inappropriate bias [35]. The second source of error stems from the fact that only a finite amount of (possibly noisy) data is available. Thus, even if we pick appropriate loss functions, models and out-of-sample estimates, the method may still yield inappropriate results [36]. The third source of error stems from the difficulty of the search problem that underlies the modeling problem under consideration. Reducing the problem to a convex optimization by appropriate choices of loss and constraints or relaxations can greatly help the search problem [33].

As far as ML algorithms are concerned, these can be distinguished in four categories concerning data classification, value prediction, structure discovery, and detection of anomalies or abnormal behavior. More specifically:

- Data classification aims to predict which category the input data belongs to. For example, in a software development project, a new task can be classified into distinct categories (e.g. story, bug, epic) based on its attributes using a decision tree classifier.
- Value prediction concerns regression algorithms to predict continuous numerical values. For example, in a common PM scenario, these algorithms can estimate the budget of a project by exploiting knowledge of similar, already accomplished projects using simple linear regression techniques, thus providing advice to the project manager during the planning phase on possible cost reduction decisions.
- Anomaly detection algorithms aim to identify unusual events or patterns that do not conform to usual or expected behavior. For example, in a certain maintenance setting, these algorithms can detect outages of some components before they occur and proactively act towards keeping the whole system functioning.
- Structure discovery aims to uncover data patterns, reveal hidden or not obvious relationships and divide data items into groups with similar traits (features). This is

achieved using widely-adopted ML techniques (e.g. k-means and Apriori algorithms). For example, in a construction PM problem, the Apriori algorithm can mine frequent itemsets concerning constructors and project durations to build useful association rules (e.g. constructor x is always late when delivering dam construction projects).

Considering the pros and cons of the techniques discussed in the previous section, in this paper we propose a hybrid approach to handle PM issues, which builds on a proper integration and orchestration of PM tools originally developed within the ML and OR disciplines. ML, which has become a buzzword nowadays [31], adopts a predictive analytics approach of the form 'if A happens, then B is likely to happen', which attempts to exploit available past data to create useful insights (i.e. make human-like decisions). On the other hand, OR adopts a prescriptive analytics approach to provide optimal solutions (courses of action) to problems of the form 'what does A need to be if we want B to happen' (i.e. make perfect decisions) [13].

We consider tools coming from the ML and OR fields as complementary, arguing that there is room for integration in a way that ML can create and refine $A \rightarrow B$ relationships that are often considered as optimal and remain unchanged upon the entry of new data in classical OR approaches. Despite the features that ML possesses in terms of data refinement and value prediction, it lacks algorithms aiming to provide optimal solutions, something that is inherent in OR techniques. Overall, our approach considers OR and ML as complementary to each other, and proposes an iterative interplay between them, where ML supplies OR algorithms with refined, accurate and up-to-date data (based on past records), while OR contributes to making optimal decisions with the continuously updated data input.

The proposed approach enables interpretation of big volumes of PM data to support preventive actions such as giving advice about resource assignments by identifying similar skills and expertise necessary to perform a task, make explainable recommendations about the capacity levels of certain resources based on historical performance data, and support informed decisions concerning a company's expansion to a new region or design of an efficient supply chain [20]. The proposed approach augments the overall PM decision-making process, by enabling the drawing of reliable conclusions about conditions and future events, while also identifying potential risks and opportunities.

4   Examples

As highlighted in the previous section, depending on the specific PM issue under consideration, our approach advocates a proper streamlining of ML and OR algorithms. In this section, we demonstrate its applicability through two realistic examples concerning resource assignment. Emphasis is given to the complementarity of ML and OR algorithms to advance the associated PM practice.

4.1   Example 1

Based on real data[1] concerning implementation of public construction projects in the Region of Attica, Greece, for the period 2003–2014, we consider the following problem [20]:

Let $P = \{P_1, P_2, ..., P_n\}$ be a set of future public construction projects. Each project (P_n) is described by a list of attributes, namely $P_n = [PID, \{M_i\}, category, est\_cost, funding, duration]$, corresponding to a unique project identifier, the municipality to manage the project, the type of construction needed, the project's estimated cost, the source funding the project, and its estimated duration, respectively.

Similarly, let $C = \{C_1, C_2, ..., C_m\}$ be the set of registered construction companies, each of them being associated with the set of attributes $[CID, \{Location_i\}, \{Category_j\}, \{Cost\_Range_k\}, \{Duration\_Range_l\}, Cost\_Overrun, Delay]$, corresponding to a unique constructor identifier, the municipality where the constructor is active, the type of projects the constructor deals with (e.g. flood control, health infrastructure), the projects' budget category the constructor is interested in (e.g. large scale (>1,5 M€), medium scale (0,5 M€–1,5 M€), the projects' duration range (e.g. short term (<6 months), mid term (6–18 months)), the project's final budget overrun/underrun, and the delay (percentage) caused by the constructor, respectively. The abovementioned data attributes are summarized in Table 1.

Table 1. Data attributes for Projects (P_n) and Constructors (C_m).

Attribute	Description	Values	Attribute type
Location	The municipality to manage the project	$\{M_i\}$	Nominal
Category	The type of the project	e.g. {Buildings, Roadworks, Athletics, ...}	Nominal
Estimated_Cost	Initial budget available	[3.000, 16.7 M] €	Numerical
Cost_Range	The project's budget category	{Small_Scale, Medium_Scale, Large_Scale}	Ordinal
Cost_overrun	The project's final budget overrun/underrun	[-76.18%, +38.55%]	Numerical (%)
Funding	The source of funding	{Region, EU, Third Party}	Nominal
Duration	The estimated duration of the project	[5, 2587] days	Numerical
Duration_Range	The project's duration range	{Short Term, Mid Term, Long Term}	Ordinal
Delay	The delay in project's accomplishment	[-402, 1485] days	Numerical

---

[1] Data available in: https://drive.google.com/file/d/13oixL7QuKtE2NidtBJEEHaoFpvylRXlH/view?usp=sharing.

Data Pre-processing. In the dataset under consideration, there exist a number of project categories containing a limited number of occurrences with respect to the total projects accomplished, as shown in Table 2. To retain only data that contribute in gaining statically significant information, we remove all records with relative project categories frequency below 1%. Moreover, constructors with limited participation in projects' construction are also removed. As a result, the original dataset is reduced from 684 to 607 transaction records.

Table 2. Relative frequencies of project categories.

Project category	Relative frequency
Athletics	2,50%
Buildings	15,02%
Cultural	**0,15%**
Drains	**0,59%**
Education infrastructure	1,91%
Flood control	16,20%
Health infrastructures	2,95%
Nursery schools	**0,44%**
Other interventions	6,92%
Planning study	5,45%
Port works	1,62%
Roadworks	35,05%
Social infrastructures	1,47%
Urban reconstruction	8,84%
Water supply	**0,88%**
Total	100%

Next, to determine whether a statistically significant difference between *Constructors$_m$* and projects' *Delay* attribute exists, we perform the Kruskal-Wallis test[2] using SPSS in the remaining records (see Table 3). The outcome of the Kruskal-Wallis test suggests that the null hypothesis should be retained, hence the *Delay* variable is excluded from our dataset.

Table 3. Kruskal-Wallis test for *Delay*.

Null hypothesis	Test	Sig.	Decision
The distribution of Delay % is the same across Constructors	Kruskal-Wallis test	0.289	**Retain the null hypothesis**

Additionally, we perform a Kruskal-Wallis test to determine whether a statistically significant difference between *Constructors$_m$* and *Cost_Overrun* variable exists. As

---

[2] We selected Kruskal-Wallis test due to the fact that our data are not normally distributed.

shown in Table 4, the outcome of the test suggests that the null hypothesis should be rejected, hence the *Cost_Overrun* variable is included in our dataset.

Table 4. Kruskal-Wallis test for *Cost_Overrun*.

Null hypothesis	Test	Sig.	Decision
The distribution of Cost_Overrun % is the same across Constructors	Kruskal-Wallis test	0.000	**Reject the null hypothesis**

Application Scenario. Let a project management scenario where there are $n = 3$ projects of various categories and $m = 6$ available constructors. Obviously, each P_n requires a different expertise, while each C_m possesses a distinct number of skills extracted from past data. To determine the constructors that best fit to the projects' requirements, we need to populate a (P_n, C_m) score matrix (each entry taking values in the range [0, 1]). This is through the calculation of: (i) the Jaccard similarity index $J(P_n, C_m)$ [19], and (ii) an additional score value $Score_{C,M}$ for the attribute *CostOverrun* of each C_m (this attribute does not participate in the calculation of the Jaccard similarity index).

We define:

$$Score(C_m, P_n) = norm[avg(cost\_overrun(C_m, P_n))] \tag{1}$$

where: $norm[avg(cost\_overrun(Cm, Pn))]$ responds to the normalized average cost_overrun for C_m constructor in the specific P_n project category.

$$J(P_n, C_m) = |P_n \cap C_m| / |P_n \cap C_m| \tag{2}$$

where: $J(P_n, C_m)$ represents the Jaccard similarity index between C_m constructor and the corresponding P_n project category.

For the calculation of the constructor's final rating, we also normalized the output values of the Jaccard similarity index in the range [0, 1]. The final rating of the C_m constructor for the P_n project is calculated by the following formula:

$$Rating_{n,m} = [a * J(P_n, C_m) - b * Score(C_m, P_n)] \tag{3}$$

where: a and b are two coefficients (in the range [0..1]) aiming to promote the best combination of similarity and cost overrun. In the specific dataset, their values are 0.9 and 0.1, respectively.

Table 5. The (P_n, C_m) score matrix $(Rating_{n,m})$.

	Athletics	Roadworks	Buildings
C_2	**0.93**	0.00	0.00
C_3	0.00	0.00	0.15
C_9	0.00	0.00	**0.40**
C_4	0.00	0.00	0.19
C_{11}	0.00	**0.44**	0.03
C_{15}	0.00	0.24	0.06

By using formulas 1–3, we calculate the (P_n, C_m) score matrix (Table 5).

Aiming to minimize the total construction cost of these projects, the problem is considered as a typical linear assignment problem (LAP), which can be easily solved through tools available in widely used software packages such as Google OR-Tools (https://developers.google.com/optimization/assignment/simple_assignment). Using the linear assignment solver of the above software package, we get the outcome presented in Table 6.

Table 6. (Pn, C_m) assignment matrix.

Project	Athletics	Roadworks	Buildings
Constructor	C_2	C_{11}	C_9

Aiming to further improve the accuracy of our estimations, we next consider the exploitation of ML algorithms, which are capable to provide knowledge-based patterns of construction projects' data. More specifically, we propose the use of the Apriori Algorithm [3] in order to discover meaningful patterns (itemsets) relating P_n and C_m attributes.

We consider the transaction set $T = \{T_1, T_2, ..., T_{607}\}$ from a total of 607 transactions available in our dataset. The application of Apriori algorithm provides us with a "strong" supported i-itemset that has been generated for constructor C_m (see Table 7; it is noted that, due to space limitations, we present only the final step of the algorithm, omitting intermediate calculations of k-itemsets). As we notice, the Apriori algorithm confirms the proposed assignment presented in Table 6 by generating the corresponding L_4 itemsets (rules), which also contain the constructors C_2, C_{11}, C_9. Apart from the above outcomes, there is a strong indication suggested by the $L_2 \cup L_4$ that in case of *Long_Term* projects, constructor C_{15} might outperform constructor C_{11}. Comparatively, the OR and ML method produce the same results; however, the ML method does not necessitate the construction of a new variable (i.e. $Rating_{n,m}$). In addition, the ML method yields a more elaborated (multi-dimensional) picture of the reality without any further configuration. In the example shown above, this concerns indication of additional constructors that may be assigned to a specific project.

Table 7. Li itemsets for constructor C_m.

Constructor (C_m)	Large itemset	(L_i)
C_2	Athletics, Small_Scale, Short_Term, Region	(L_4)
C_{11}	Roadworks, Small_Scale, Short_Term, Region	(L_4)
C_9	Buildings, Small_Scale, Short_Term, Region	(L_4)
C_{15}	**Roadworks, Small_Scale, Mid_Term, Region**	**(L_4)**
C_{10}	Flood Control, Small_Scale, Short_Term	(L_3)
C_{15}	**Long_Term, Region**	**(L_2)**

Our approach is sketched in a pseudo-code form below:

Algorithm 1.1

```
1: for each (Pn) do
       {
2:        find similarity(Pm, Cm);
3:        calculate_Score(Pn, Cm) matrix;
       }
4: assign(Pn, Cm);
5: for each Cn do
       {
6:        Apply Apriori_Algorithm;
7:        Generate Li-itemsets;
       }
8: assign(Pn, Cm);
```

To summarize the basic concepts of the above example, we addressed a PM issue as a typical OR assignment problem (a group of constructing companies need to accomplish a set of construction projects) using a score matrix with estimations for each *(Pn, Cm)* element. The LAP solver algorithm provided a solution to the problem prescribing the optimal assignment matrix. Next, we exploited ML (the Apriori algorithm) to discover association rules between transactions' data to spot trends, relationships and structure similarity between data sets. In this way, we demonstrated that ML models and algorithms can be used to re-feed/alter initial OR solutions, integrating OR's prescriptive analytics with ML's predictive analytics orientation.

4.2   Example 2

In this example[3], we extract and analyze data[4] from the publicly accessible Jira instance of Apache Software Foundation (https://issues.apache.org/jira). This dataset concerns the development (i.e. issue tracking, bug fixing, implementation of new features, etc.) of the Apache Hadoop project [26] and contains information related to 1000 Jira issues. It has been retrieved using the open-source Python library 'jira' (https://github.com/pycontribs/jira), which relies on the official REST API of Jira. Each Jira issue in our dataset has, among others, the following important attributes (Table 8):

Table 8. Data attributes for Jira issues.

Attribute	Description	Values	Attribute type
Id	The identifier of the issue	e.g. {1345}	Integer
Key	The textual identifier of the issue	e.g. {HADOOP-1345}	String
Labels	The labels of the issue	e.g. {KeyStore, security, tpm}	String
Assignee	The assignee of the issue	e.g. {john_doe, jane_doe}	String
Status	The project's current status	{Close, In Progress, Open, Patch Available, Reopened, Resolved}	String
Components	The parental architectural components of the module that concerns the issue	e.g. {Back-end, Front-end, Main-Framework}	String
Description	The description of the issue	Unstructured text	String
Summary	The title of the issue	Unstructured text	String
Reporter	The reporter of the issue	e.g. {john_doe, jane_doe}	String
Resolution Date	The resolution date of the issue	e.g. {1560771216}	Timestamp
Created at	The date the issue has been created	e.g. {1560771216}	Timestamp

Data Cleansing. Similarly to the previous example, we retain only the Jira issues that: (i) have an assignee (in our dataset 677 out of 1000), and (ii) contribute in gaining statistically significant information (174 out of 677). As resulted, most of the remaining Jira issues have been assigned to only 4 developers (see Fig. 1 - for obvious reasons, the real names of the developers have been replaced by common first names).

---

[3] Code available at: https://github.com/nkanak/advance-project-management-practice.

[4] Data available at: https://github.com/nkanak/advance-project-management-practice/blob/master/data/hadoop_issues.json, retrieved at: 10 Dec 2018.

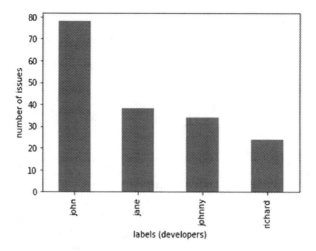

Fig. 1. Number of issues assigned per developer.

Feature Generation and Feature Extraction. To extract features needed in the next steps, a variety of Natural Language Processing (NLP) techniques, including feature generation and text normalization, is applied to the most important textual attributes of each issue, namely 'summary' and 'description'. This process can be accomplished through the following steps:

- Tokenization: the process of generating tokens from unstructured text;
- Lemmatization: the process of grouping together the different inflected forms of a word;
- Stemming: the process of reducing derived words to their root form;
- Feature/Tag generation: the process of transforming an unstructured text into a vector that contains word occurrences;
- Removal of stop words: the process of removing commonly used words, which usually add noise to the ML models;
- Removal of frequently occurred words: the process of removing insignificant words from a text, taking into consideration the document frequency value of each word; in our example, we remove the words with document frequency greater than 60%.

Application Scenario. We aim to increase the success chances of the project by properly assigning the available developers to Jira issues. We argue that a project has more chances to succeed if this assignment considers the skills of each developer and the nature of the job to be accomplished in each issue. For a given issue X and a developer Y, we define a relevance metric as the probability that the developer Y possesses the skills required by issue X and, consequently, his/her capability of successfully completing the issue on time [16]. Our approach includes the following two steps:

- The estimation of how relevant each developer is to undertake (and successfully accomplish) each Jira issue;
- The assignment of developers to tasks, based on the linear assignment problem (LAP) algorithm [6], in a way that maximizes the total relevance metric.

As far as the calculation of relevance is concerned, we train a ML model (number of training records: 118, number of testing records: 56) that exploits the available textual information concerning each issue (i.e. 'summary' and 'description' attributes). Hence, we reduce the relevance calculation problem to the common and well-studied text classification problem. We adopt the multinomial Naive Bayes[5] classifier since it is suitable for classification with discrete features such as word counts [2]. The textual information of each issue is assigned to a class (i.e. the class with the highest probability) [25]. The set of classes of our model consists of the names of the available developers (i.e. 'assignee' attribute). Despite the fact that the multinomial Naive Bayes classifier assigns only one class (i.e. developer) to an issue, it also predicts the probability that the issue belongs to all other classes (i.e. how relevant is this issue to each available developer). It is clear that in our approach the relevance metric for each developer is equivalent to the probability calculated by our model.

For the needs of the example described in this section, we elaborate an application scenario using the textual information contained in the following four issues of our dataset:

Issue 1

Text: ``branch-2 site not building after ADL troubleshooting doc added
Toc error on the ADL troubleshooting doc from HADOOP-15090{code}[ERROR]
Failed to execute goal org.apache.maven.plugins:maven-site-
plugin:3.5:site (default-cli) on project hadoop-azure-datalake: Error
parsing 'hadoop-trunk/hadoop-tools/hadoop-azure-
datalake/src/site/markdown/troubleshooting_adl.md': line [-1] Error
parsing the model: Unable to execute macro in the document: toc -> [Help
1]{code}''
Actual Class: ``john''
Predicted Class: ``john''

---

[5] Python class used: `sklearn.naive_bayes.MultinomialNB`.

Issue 2

Text: ``S3 listing inconsistency can raise NPE in globber FileSystem Globber does a listStatus(path) and then, if only one element is returned, {{getFileStatus(path).isDirectory()}} to see if it is a dir. The way getFileStatus() is wrapped, IOEs are downgraded to null. On S3, if the path has had entries deleted, the listing may include files which are no longer there, so the getFileStatus(path),isDirectory triggers an NPE. While its wrong to glob against S3 when its being inconsistent, we should at least fail gracefully here.
Proposed
log all IOEs raised in Globber.getFileStatus @ debug
catch FNFEs and downgrade to warn
continue
The alternative would be fail fast on FNFE, but that's more traumatic''
Actual Class: ``john''
Predicted Class: ``john''

Issue 3

Text: ``ABFS: Code changes for bug fix and new tests
- add bug fixes.
- remove unnecessary dependencies.
- add new tests for code changes.''
Actual Class: ``johnny''
Predicted Class: ``johnny''

Issue 4

Text: ``Release Hadoop 2.7.7
Time to get a new Hadoop 2.7.x out the door.''
Actual Class: ``john''
Predicted Class: ``john''

Applying the LAP algorithm[6] to the elements of Table 9, the final step assigns developers to issues.

Table 9. Relevance of each developer per issue/task (%).

	T_0	T_1	T_2	T_3
John	96	95	8	53
Jane	0	1	0	43
Richard	4	0	0	3
Johnny	0	4	92	1

---

[6] https://developers.google.com/optimization/assignment/simple_assignment.

The outcome of the above process is shown in Table 10.

Table 10. The issue/task assignment matrix.

Issue	T_0	T_1	T_2	T_3
Developer	Richard	John	Johnny	Jane

Evaluation Measures. The above multinomial Naive Bayes classifier has a mean accuracy score[7] of 82.7%. In order to get a solid understanding of the underlying mechanism of our trained model, we explain the predictions of the model using the *Local Surrogate Models (LIME)* explanation method [15, 32]. In brief, this method tries to explain why single predictions of black-box ML classifiers were made by perturbing the dataset and building local interpretable models. In this case, the LIME text explainer[8] randomly removes words/features from the text of each issue and calculates the importance of a specific word to the decision made by the Naive Bayes classifier.

The provided explanations help us check the [21] of the trained ML model; they also confirm that the model selects the right label/class for the right reason (i.e. meaningful words/features). For instance, Fig. 2 explains why 'john' is (correctly and rationally) selected to work on a specific Jira issue with the following text:

Text: "branch-2 site not building after ADL troubleshooting doc added.
Toc error on the ADL troubleshooting doc from HADOOP-15090 {code}
[ERROR] Failed to execute goal org.apache.maven.plugins:maven-site-
plugin:3.5:site (default-cli) on project hadoop-azure-datalake: Error
parsing 'hadoop-trunk/hadoop-tools/hadoop-azure-
datalake/src/site/markdown/troubleshooting_adl.md': line [-1] Error
parsing the model: Unable to execute macro in the document: toc -> [Help
1]{code}".

As illustrated in Fig. 2, features such as *'hadoop'*, *'adl'*, *'doc'*, *'azure'* and *'troubleshoot'* play an important role in the decision made (choosing 'john') by the classifier. Additionally, features such as *'maven'*, *'tools'*, *'plugin'*, *'error'* and *'project'* increase the probability of 'richard' being the most suitable developer for the selected issue.

---

[7] Python method used: `sklearn.naive_bayes.MultinomailNB.score`.

[8] Python class used: `lime.lime_text.LimeTextExplainer`.

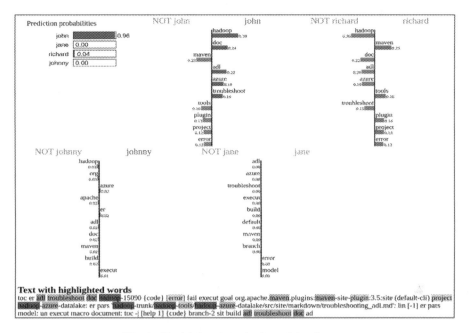

Fig. 2. Explaining the selection of developers.

Future Improvements. This example highlights the need for leveraging ML and OR techniques to augment the outcome of the classic linear assignment problem. However, during the development of our approach, various limitations and problems have been identified. First of all, the amounts of statistically significant data are limited. Secondly, the developed classifier, in some cases, predicts the correct developer for wrong reasons, as depicted in Fig. 3; features such as *'door'*, *'new'*, *'release'* and *'time'* affect the decision of our ML model. Finally, several important attributes remain unused; these attributes include the resolution time and the importance of an issue (e.g. *'blocker'*, *'critical'*, *'major'*, *'minor'*, *'trivial'*).

To overcome the abovementioned shortcomings, we aim to enrich our approach; as far as the ML part of our approach is concerned, further removal of stop words, synonym identification [4], and part-of-speech tagging [24] is required. Also, as proposed in [38], the enrichment of the corpus using Wikipedia knowledge may improve the accuracy of the Naive Bayes classifier. Furthermore, contemporary knowledge representations such as knowledge graphs [5] and 'document to graph' [29, 41, 42] may mitigate problems that have arisen from the 'curse-of-dimensionality' phenomenon and improve the accuracy of the classifier. In the OR part of our approach, constraint satisfaction solvers can be integrated in order to enable the exploitation of usable features ranging from time and budget constraints to scheduling problems (e.g. top-ranking issues with the highest importance) [18].

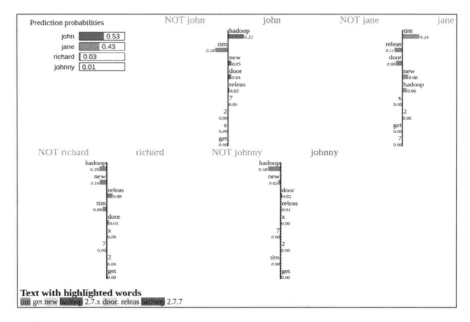

Fig. 3. Wrong classifier predictions.

5 Discussion

Key enablers that are driving the development of the proposed approach are the availability of huge computing power, the existence of big volumes of PM data and knowledge, as well as the accessibility of a range of well-tried and powerful OR and ML software libraries. Undoubtfully, there is more computing power available today than ever before, something that contributes significantly in making OR and ML algorithms extremely powerful, in ways that were not possible even a few years ago. In fact, this computing power enables us today to process massive amounts of PM data and extract valuable knowledge needed to make our models more intelligent. At the same time, as discussed in Sect. 2, software needed to process the diversity of PM data is open and freely available; it is also noted here that PM-related AI algorithms become available and get commoditized via dedicated APIs (Application Programming Interfaces) and cloud platforms.

Despite the above advancements, much work must still be carried out on the proper manipulation of PM data and knowledge, as far as its labeling, interrelation, modeling and assessment are concerned; this has mainly to be done by humans. Especially in the context of project management, one should always take into account that valuable data and knowledge emerge continuously during an organization's lifecycle, and concern both the organization *per se* (e.g. a project's duration, overall budget, KPIs etc.) and its employees (e.g. one's competences and performance, knowledge shared during a decision-making process etc.).

Building on a meaningful and flexible integration of OR and ML techniques and associated tools, our approach enables organizations to reap the benefits of the AI revolution. It allows for new working practices that may convert information overload and cognitive complexity to a benefit of knowledge discovery. This is achieved through properly structured data that can be used as the basis for more informed decisions. Simply put, our approach improves the quality of PM practice, while enabling users to be more productive and focus on creative activities. However, diverse problems and limitations still exist; these concern the value and veracity of existing data, as well as the availability of the massive amounts of data required to drive contemporary AI approaches.

6   Conclusions

This paper presents a hybrid approach aiming to assist the overall PM practice. We demonstrated that in ML and OR exist a variety of techniques enhancing a vast number of issues and tasks such as: resource assignment problems, task(s) duration estimation and task accomplishment prediction. Optimization and big data manipulation both a key issue in ML and OR correspondingly, are handled with equal attention in order to reach the desired outcome in PM tasks. It is worth to notice that our work further extends [20] by embedding explainability features in the recommendations provided in the presented examples.

From a broader point of view, we also argue that there is a major lack of scientific research that relates to AI solutions being developed for specific functions such as project management [30]. Thus, a key issue for future research is to study AI-based solutions for specific functions purely from a scientific perspective. On the other hand, there are no scientific theories that help one understand how a technology that has not been fully understood or developed yet can affect functions that traditionally rely on cognitive input or human interactions [36]. Therefore, another research direction should contribute to the development of theoretical models in order to help understand the impact of technologies on such functions.

References

1. Abadi, M., et al.: Tensorflow: a system for large-scale machine learning. In: Proceedings of the 12th USENIX Symposium on Operating Systems Design and Implementation, Savannah, pp. 265–283 (2016)
2. Aggarwal, C.C., Zhai, C.: Mining Text Data. Springer, Heidelberg (2012). https://doi.org/10.1007/978-1-4614-3223-4
3. Agrawal, R., Srikant, R.: Fast algorithms for mining association rules. In: Proceedings of 20th International Conference on Very Large Data Bases (VLDB 1215), San Francisco, pp. 487–499 (1994)
4. Alkhraisat, H.: Issue tracking system based on ontology and semantic similarity computation. Int. J. Adv. Comput. Sci. Appl. 7(11), 248–251 (2016)
5. Betts, C., Power, J., Ammar, W.: GrapAL: Querying Semantic Scholar's Literature Graph. arXiv preprint arXiv:1902.05170 (2019, to appear)

6. Burkard, R.E., Dell'Amico, M., Martello, S.: Assignment Problems, Philadelphia (2009)
7. Burke, L.I., Ignizio, J.P.: Neural networks and operations research: an overview. Comput. Oper. Res. **19**(3), 179–189 (1992)
8. Candel, A., Parmar, V., LeDell, E., Arora, A.: Deep Learning with H2O, 6th edn. H2O.ai Inc. http://h2o.ai/resources/. Accessed 19 Oct 2018
9. Cheung, S.O., Wong, P.S.P., Fung, A.S., Coffey, W.: Predicting project performance through neural networks. Int. J. Project Manag. **24**(3), 207–215 (2006)
10. Costantino, F., Gravio, G.D., Nonino, F.: Project selection in project portfolio management: an artificial neural network model based on critical success factors. Int. J. Proj. Manag. **33** (8), 1744–1754 (2015)
11. Dolan, E., More, J.: Benchmarking optimization software with performance profiles. Math. Program. **91**(2), 201–213 (2002)
12. Dunning, I., Huchette, J., Lubin, M.: Jump: a modeling language for mathematical optimization. SIAM Rev. **59**(2), 295–320 (2017)
13. Evans, J.R., Lindner, C.H.: Business analytics: the next frontier for decision sciences. Decis. Line **43**(2), 4–6 (2012)
14. Goodfellow, I., Bengio, Y., Courville, A.: Deep Learning. The MIT Press, Cambridge (2016)
15. Guidotti, R., Monreale, A., Ruggieri, S., Turini, F., Giannotti, F., Pedreschi, D.: A survey of methods for explaining black box models. ACM Comput. Surv. (CSUR) **51**(5), 93 (2018)
16. Hill, J., Thomas, L.C., Allen, D.B.: Experts' estimates of task durations in software development projects. Int. J. Proj. Manag. **18**(1), 13–21 (2000)
17. Holmes, G., Donkin, A., Witten, I.H.: Weka: a machine learning workbench. In: Proceedings of the 1994 Second Australian and New Zealand Conference, Adelaide, pp. 357–361 (1994)
18. Hooker, J.N., Van Hoeve, W.J.: Constraint programming and operations research. Constraints **23**(2), 172–195 (2018)
19. Jaccard, P.: Étude comparative de la distribuition florale dans une portion des Alpes et des Jura. Bull. Soc. Vandoise Sci. Nat. **37**, 547–579 (1901)
20. Kanakaris, N., Karacapilidis, N., Lazanas, A.: On the advancement of project management through a flexible integration of machine learning and operations research tools. In: 8th International Conference on Operations Research and Enterprise Systems (ICORES), Prague, pp. 362–369 (2019)
21. Karacapilidis, N., Malefaki, S., Charissiadis, A.: A novel framework for augmenting the quality of explanations in recommender systems. Intell. Decis. Technol. J. **11**(2), 187–197 (2017)
22. Karacapilidis, N.: Mastering Data-Intensive Collaboration and Decision Making: Cutting-Edge Research and Practical Applications in the Dicode Project. Studies in Big Data Series, vol. 5. Springer, Cham (2014). https://doi.org/10.1007/978-3-319-02612-1
23. Levitt, R.E., Kunz, J.C.: Using artificial intelligence techniques to support project management. Artif. Intell. Eng. Des. Anal. Manuf. **1**(1), 3–24 (1987)
24. Maurya, A., Telang, R.: Bayesian multi-view models for member-job matching and personalized skill recommendations. In: 2017 IEEE International Conference on Big Data, pp. 1193–1202. IEEE (2017)
25. Mooney, R.J., Roy, L.: Content-based book recommending using learning for text categorization. In: Proceedings of the fifth ACM conference on Digital libraries, pp. 195–204. ACM (2000)
26. O'Malley, O.: Terabyte sort on apache hadoop. Yahoo, pp. 1–3 (2008). http://sortbenchmark.org/Yahoo-Hadoop.pdf

27. Paszke, A., Gross, S., Chintala, S., Chanan, G., Yang, E., DeVito, Z., Lerer, A.: Automatic differentiation in pytorch. In: 31st Conference on Neural Information Processing Systems, Long Beach, pp. 1–4 (2017)
28. Pedregosa, F., et al.: Scikit-learn: machine learning in Python. J. Mach. Learn. Res. **12**(1), 2825–2830 (2011)
29. Pittaras, N., Giannakopoulos, G., Tsekouras, L., Varlamis, I.: Document clustering as a record linkage problem. In: Proceedings of the ACM Symposium on Document Engineering, p. 39. ACM (2018)
30. Radin, R.L.: Optimization in Operations Research. Prentice-Hall, New Jersey (1998)
31. Raschka, S.: Python Machine Learning. Packt Publishing Ltd., Birmingham (2015)
32. Ribeiro, M.T., Singh, S., Guestrin, C.: Why should I trust you? Explaining the predictions of any classifier. In: Proceedings of the 22nd ACM SIGKDD International Conference on Knowledge Discovery and Data Mining, pp. 1135–1144. ACM (2016)
33. Rifkin, R., Klautau, A.: In defense of one-vs-all classification. J. Mach. Learn. Res. **5**(1), 101–141 (2004)
34. Rousu, J., Saunders, C., Szedmak, S., Shawe-Taylor, J.: Kernel-based learning of hierarchical multilabel classification models. J. Mach. Learn. Res. **7**(1), 1601–1626 (2006)
35. Rummelhart, D., Hinton, G., Williams, R.: Learning Internal Representations by Error Propagation. Parallel Distributed Processing. MIT Press, Cambridge (1986)
36. Shivaswamy, P.K., Bhattacharyya, C., Smola, A.J.: Second order cone programming approaches for handling missing and uncertain data. J. Mach. Learn. Res. **7**(1), 1283–1314 (2006)
37. Smith, K.A., Gupta, J.N.: Neural networks in business: techniques and applications for the operations researcher. Comput. Oper. Res. **27**(11), 1023–1044 (2000)
38. Spanakis, G., Siolas, G., Stafylopatis, A.: Exploiting Wikipedia knowledge for conceptual hierarchical clustering of documents. Comput. J. **55**(3), 299–312 (2012)
39. Svejvig, P., Andersen, P.: Rethinking project management: a structured literature review with a critical look at the brave new world. Int. J. Proj. Manag. **33**(2), 278–290 (2015)
40. Trupti, M.K., Prashant, R.M.: Review on determining number of cluster in K-means clustering. Int. J. Adv. Res. Comput. Sci. Manag. Stud. **1**(6), 90–95 (2013)
41. Tsekouras, L., Varlamis, I., Giannakopoulos, G.: Graph-based text similarity measure that employs named entity information. In: RANLP, pp. 765–771 (2017)
42. Vazirgiannis, M., Malliaros, F.D., Nikolentzos, G.: GraphRep: boosting text mining, NLP and information retrieval with graphs. In: Proceedings of the 27th ACM International Conference on Information and Knowledge Management, pp. 2295–2296. ACM (2018)
43. Wang, Y.R., Yu, C.Y., Chan, H.H.: Predicting construction cost and schedule success using artificial neural networks ensemble and support vector machines classification models. Int. J. Proj. Manag. **30**(4), 470–478 (2012)
44. Zhang, G., Patuwo, B.E., Hu, M.Y.: Forecasting with artificial neural networks: the state of the art. Int. J. Forecast. **14**(1), 35–62 (1998)

Metaheuristics for Periodic Electric Vehicle Routing Problem

Tayeb Oulad Kouider, Wahiba Ramdane Cherif-Khettaf[(⊠)],
and Ammar Oulamara

Université de Lorraine, Lorraine Research Laboratory in Computer Science
and its Applications, LORIA (UMR 7503), Campus Scientifique,
615 Rue du Jardin Botanique, 54506 Vandœuvre-les-Nancy, France
{tayeb.ouladkouider,ramdanec,oulamara}@loria.fr

Abstract. This paper proposes two metaheuristics based on large neighbourhood search for the PEVRP (Periodic Electric Vehicle Routing Problem). In the PEVRP a set of customers have to be visited, one times, on a given planning horizon. A list of possible visiting dates is associated with each customer and a fixed fleet of vehicles is available every day of the planning horizon. Solving the problem requires assigning a visiting date to each customer and defining the routes of the vehicles in each day of the planning horizon, such that the EVs could be charged during their trips at the depot and in the available external charging stations. The objective of the PEVRP is to minimize the total cost of routing and charging over the time horizon. The first proposed metaheuristic is a Large Neighbourhood Search, whose choice of destroy/repair operators has been determined according to the experimental results obtained in previous research. The second method is an Adaptive Large Neighborhood Search, which could be described as a Large Neighborhood Search algorithm with an adaptive layer, where a set of three destroy operators and three repair operators compete to modify the current solution in each iteration of the algorithm. The results show that LNS is very competitive compared to ALNS for which the adaptive aspect has not made it more competitive than the LNS.

Keywords: Periodic vehicle routing · Electric vehicle · Charging station · Large Neighborhood Search · Adaptive Large Neighborhood Search

1   Introduction

To reduce carbon emissions in city centers, the electric vehicle has emerged as a credible alternative solution and several companies in different sectors have started to use EVs in their operations, but the use of EVs in logistics still faces several technical challenges. These challenges are related to the availability of electric vehicles with an appropriate volume and capacity load for the large-scale use, the limited driving range, long charging time, and the availability of a

© Springer Nature Switzerland AG 2020
G. H. Parlier et al. (Eds.): ICORES 2019, CCIS 1162, pp. 156–170, 2020.
https://doi.org/10.1007/978-3-030-37584-3_8

charging infrastructure. To overcome these challenges, considerable research has been devoted to develop optimization techniques for solving the electric vehicle routing problems (EVRPs).

In this paper, we consider distribution systems, in which the planning horizon is not on a daily basis, rather there is a periodicity in deliveries so that customers must be served several times in the horizon according to a frequency and a feasible pattern of delivery days defined for these customers. This problem is known as the periodic vehicle routing problems (PVRP). The aim of the PVRP is to design the arrangement of the customer's deliveries in a given period as a set of routes, one per vehicle, for every day of the planning horizon such as the total distance travelled must be minimized.

The Periodic Vehicle Routing Problems (PVRP) has received much attention in the literature, since it arises in many real-world applications such as the routing of healthcare nurses, the transportation of elderly or disabled persons, and the delivery of groceries. It extends the basic VRP to a planning time horizon of several days [5]. The objective of the PVRP is to find an optimal schedule of customer's deliveries that minimizes total travel time over time horizon of h periods of days while satisfying vehicle capacity, predetermined visit frequency for each client, and spacing constraints. As the PVRP is NP-hard problem, heuristics and metaheuristics have been widely used to solve it compared to exact methods [1, 3, 6, 17]. A survey on the PVRP can be found in [9].

A promising extension of PVRP models is the integration of constraints related to the use of electric vehicles. Indeed, as cited above the electric vehicle represents a viable solution to meet with the new policies and regulations related to greenhouse gas emission in the transport sector. Several real applications in the transportation sector previously modelled by PVRP are concerned with the necessity to use electric vehicles. A significant number of papers on several variants of EVRP have been published in recent years. As an example, we can cite the EVRP with time windows and recharging stations presented in [25]; EVRP problem with mixed fleet of electric and conventional vehicles and time windows constraints [11], EVRP with heterogeneous electric vehicles constraints [13], EVRP with different charging technologies and partial EV charging [8], the rich variant of EVRP [21, 22], a two-echelon EVRP [14], electric vehicles routing and locating charging stations [24] and [23], EVRP with pickup and delivery with time windows [12]. The most recent survey on the EVRP is presented by [7, 18] and [18].

Even with the growing interest in EVRPs, the consideration of multi-period planning has received very little attention in the literature. The PEVRP was studied in [15, 16]. The authors propose to deal with tactical and operational decisions level for electric vehicles routing and charging in which the frequency visit was fixed to one. Constructive heuristics was proposed to solve the problem in [16], The result was improved with Large Neighbourhood Search (LNS) in [15].

The contribution of this article compared to the study presented in [15] is (1) the investigation of another metaheuristic named Adaptive Large Neighbourhood Search (ALNS) for the PEVRP, (2) a comparative study between LNS and

ALNS to determine if our LNS in which the choice of operators was optimized remains competitive compared to a classical framework of ALNS, where the choice of operators is completed in an adaptive way.

The rest of the paper is organized as follows. Section 2 gives more details on constraints and characteristics of our problem. Section 3 describes our solving approaches based on LNS and ALNS. Section 4 presents experimental tests. Section 5 concludes this study.

2   Problem Definition

This study focuses on the Periodic Electric Vehicle Routing Problem (PEVRP), which has already been introduced in [16] and [15]. We only list here the important data of the problem. For more details on these data, we refer the reader to [15].

The PEVRP is defined on a time horizon H of np periods typically "days", in which each customer i has a frequency $f(i) = 1$ of visits. This means that customer i must be serviced one time over the planning horizon but at most once per day. Each customer i has a set of allowed visit days $D(i) \in H$. The PEVRP is modelled by a complete directed graph $G = (V, A)$. $V = C \cup B \cup \{0\}$, where the set C of n vertices represents the customers, the set B of ns vertices denotes the external charging stations, which can be visited during each day of the planning horizon, and the vertex 0 represents the depot where the vehicles are based. The depot contains internal charging stations allowing charging at night and during the day. A is the arcs set, where each arc (i, j) has a travel cost c_{ij}, a travel distance d_{ij}, travel time t_{ij}, and energy consumption $e_{i,j} = r \times d_{i,j}$, where r denotes a constant energy consumption rate.

The PEVRP consists in assigning each customer i to a service day in $D(i)$ that minimize the total cost of routing and charging over H. A feasible solution of PEVRP must satisfy the following set of constraints: (i) each route must start and end at the depot, (ii) each customer i should be visited one time during the planning horizon according to one day in $D(i)$, (iii) the customer demand q_i must be completely fulfilled (iv) no more than m electric vehicles are used in each day, (v) the total duration of each route, calculated as the sum of, travel duration required to visit customers, time required to charge the vehicle during the day, and the service time of each customer; could not exceed T; (vi) the overall amount of goods delivered along the route, given by the sum of demands q_i of visited customers, must not exceed the vehicle capacity Q.

The objective function to be minimized is $f(x) = \alpha \times f_1(x) + Cc \times nbs(x)$ where, $f_1(x)$ is the total distance of the solution x over the planning period H, $nbs(x)$ is the number of visits to charging stations in solution x over the planning period, Cc is a fixed charging cost, and α is a given weight representing the cost of one unit of distance.

3   Metaheuristic Approaches for the PEVRP

In [15], we proposed a resolution method based on Large Neighborhood Search (LNS). The LNS starts from a given feasible solution and improves it using destroy and repair operators. The choice of operators was optimized since a detailed analysis of the performance of three destroy and three repair operators were performed and led us to retain the best couple of operators in the proposed LNS. The purpose of this study is to compare the LNS of [15] with an adaptive approach named ALNS. The ALNS proposed by Pisinger and Røpke [20] can be described as a large neighborhood search algorithm with an adaptive layer, where the three destroy operators and the three destroy operators compete to modify the current solution in each iteration of the algorithm. In this paper, we are interested to determine if our LNS in which the choice of operators was optimized remains competitive compared to LNS where the choice of operators is completed in an adaptive way.

The rest of this section is organized as follows. In Sect. 3.1 we explain how we addressed energy constraints in our approaches, in Sects. 3.2 and 3.3 we describe the proposed destroy and repair operators, the initial solution will be depicted in Sect. 3.4, in Sect. 3.5 we briefly recall the LNS, and in Sect. 3.6 the ALNS based solution method will be detailed.

3.1   Consideration of Energy Constraints

We propose specific procedures that are necessary to deal with energy and charging constraints. These procedures are used whenever a destroy or repair operator is used to repair routes if necessary and adjust the required amount of energy.

$AdjustDecreaseCharging(Tr)$: This method is applied to a partial solution after a destroy operator was executed. After deleting some customers by a destroy operator, this procedure evaluates the unused energy of each modified route Tr (a route from the current solution in which at least one customer is deleted) and decides whether a station should be deleted and/or which station should reduce the amount of its charged energy.

$AdjustDecreaseCharging(Tr)$ receives as an input a sequence Tr. It first tries to remove useless stations and then reduces the energy provided by each station. It repeats these two above steps until the amount of energy consumed by Tr cannot be decreased any more. The Fig. 1 describes the principle of the $AdjustDecreaseCharging(Tr)$ procedure. The detailed algorithm is given in [15].

$AdjustIncreaseCharging(Tr)$: this method is used by the insertion operators. It is applied after the insertion of each customer in a given sequence Tr to estimate the energy to be injected into Tr. It decides which stations will have to increase their energy, and/or, if necessary, which new stations will be added to the tour Tr. The algorithm first tries to inject more energy into the existing stations, and if it cannot repairs the tour Tr, it tries to insert one (or more) new stations that minimize the increased cost. The Fig. 2 describes the principle of

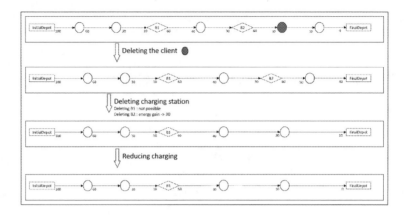

Fig. 1. *AdjustDecreaseCharging* example (source: [15]).

the *AdjustIncreaseCharging*(Tr) procedure. The detailed algorithm is given in [15]. Note that the algorithm fails to insert an ejected customer i in some cases. This can occur if (1) the customer i cannot be inserted into the existing tours because of the capacity and/or time limit constraints, and the vehicle limit is reached in each day p of the horizon, or (2) for each feasible insertion position given by a day $h \in D(i)$, a tour t scheduled in the day h, and a position $k \in t$, such that the constraint of capacity and total time are satisfied but the energy constraint is not satisfied for t: there is no charging station that can repair t.

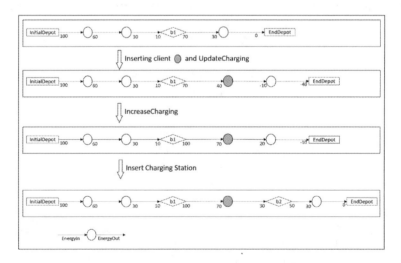

Fig. 2. *AdjustIncreaseCharging* example (source: [15]).

3.2   Destroy Operators

The destroy operators delete a given number of customers (denoted γ) from a current solution. We propose to adapt three classical removal operators from the literature [19]. They consist in the random removal, the worst removal, and the cluster removal. We detail these operators in the following. Let $S = \{S_1, .., S_{np}\}$ the initial feasible solution, and S_h a set of routes in the day h, $S_h = \{T_{1h} \ldots T_{kh}..T_{mh}\}$. Let S' the partial solution obtained after removing γ customers, $S' = \{T'_{1h} \ldots T'_{kh}, .., T'_{mh}\}$, such that $T'_{kh} = T_{kh}$ if no customer has been ejected from T_{kh}, and $T'_{kh} = T^-_{kh}$ if one or more customers has been ejected from T_{kh}.

Random Removal$(S, S', f(S'))$: This operator receives the solution S as input, first randomly deletes γ customers in any tour and on any day, then applies the repair procedure *AdjustDecreaseCharging*(Tr), and finally returns the obtained partial solution S' and its cost $f(S')$.

Worst Removal $(S, S', f(S'))$: This operator ejects the customers, whose removal produces the biggest decrease in the objective function. The idea behind this operator is to try to reinsert these clients into more advantageous positions. This procedure deletes the customers one by one until γ is reached. For each customer $i \in S$, the algorithm computes $f(S^{-i})$ which represents the cost of the solution S without the client i. In fact, the algorithm simulates the removal of i, by deleting i from S, and applying *AdjustDecreaseCharging*$(trip(i)^{-i})$, with $trip(i)^{-i}$ denotes the tour where the client i was inserted without the client i.

Cluster Removal $(S, S', f(S'))$: The notion of a cluster in our problem is related to a set of nodes close to each other in terms of solution cost. These nodes can belong to different routes and different days, which ensures that the neighbourhood considered by the cluster is large enough. This operator chooses a customer i randomly, then $\gamma - 1$ additional customers located nearest to i (in terms of costs) are selected to form the cluster to be removed from S. *AdjustDecreaseCharging*(Tr) procedure is applied to each route $T^-_{kl} \in S'$ and the new cost of the partial solution $f(S')$ is returned, with T^-_{kl} denotes the route k of a day l, where at least one customer has been deleted.

3.3   Repair Operators

We have proposed three repair operators, namely, First Improvement, Best Improvement, Regret Insertion. After each customer insertion, the procedure *AdjustIncreaseCharging*(Tr) is applied to adjust the energy of each modified tour $Tr \in S$ ($Tr \in S$ tour concerned by the removal of at least one customer). If the insertion of the customer i fails in the existing routes, a new route containing the depot and the customer i can be built if the maximum number of vehicles is not reached. When all customers have been reinserted back into the solution, the new solution is compared with the original solution. If it is impossible to insert all ejected clients, the solution may become unfeasible. For each repair operator, we have implemented two insertion strategies to define the number of positions to be tested for the insertion of a given client. The first strategy named $InsertC^{All}$ tests all insertion positions ($|Tr| - 1$ positions for a given tour Tr) and in the second strategy, named $InsertC^{\alpha}$, α nodes $u_1 \ldots u_{\alpha}$ closest to i in terms of distance are selected. The insertion positions held are then the positions before and after each selected node.

First Improvement Insertion. Let L be the list of ejected customers. This method selects randomly a node in L and try to insert it in the position that generates the minimal cost increase. If the insertion of client i into a position l of a route t fails due to capacity constraint and/or time limit, the position l is rejected, otherwise the energy constraint is checked and the procedure $AdjustIncreaseCharging(t)$ is applied if necessary.

Best Improvement Insertion. Let L be the list of ejected customers. This procedure iterates the three steps below until $L = \emptyset$. Step 1. Computes the minimum cost insertion of each customer $i \in L$ by scanning all positions in each tour Tr and in each day and uses $AdjustIncreaseCharging(Tr)$ if necessary. Step 2. The customer i^* (and eventually the charging stations b^*), Step 3. The set L and the current solution are updated.

Regret Insertion. Let L be the list of ejected customers. This procedure iterates the three steps below until $L = \emptyset$. Step 1. Computes the difference between the two best cost insertions of each customer $i \in L$, denoted δ_i and uses $AdjustIncreaseCharging(Tr)$ to repair the solution if necessary. Step 2. A customer i^* (and eventually the charging station b^*) with the maximum δ_{i^*} is inserted at its best position. Step 3. The set L and the current solution are updated.

3.4   Initial Solution

We suggest testing our methods with two initial solutions. These two solutions are named BIH (Best Insertion Heuristic) and CLH (Clustering heuristic). They have been proposed in [16]. The BIH heuristic consists of inserting each customer i (and when necessary, a charging station b) at its best position, where a position is characterized by a day $h \in D(i)$, a tour t scheduled in the day h, and a position $k \in t$. The CLH heuristic starts, in a first step, by generating m clusters in each day of the horizon ans inserts customers whose all constraints are verified, then in a second step, the heuristic considers the insertion of customers who have violated the energy constraints in the first step. The insertion criteria of these customers is the minimization of the additional energy consumption for each cluster. Finally, a best insertion TSP heuristic is used to find a feasible route in each cluster for each day.

3.5   Large Neighborhood Search Framework

Our first method approach uses the LNS framework and was proposed in [15]. The LNS starts from one initial solution (CLH or BIH) and improves it using the Destroy-Repair process. Indeed, LNS removes a relatively large number of customers from the current solution and tries to reinsert them into different positions. This leads to a completely different solution, that helps the heuristic to escape local optima. In previous study [15], we implemented nine versions of LNS. Each implemented LNS version consists of a different pair of destroy/repair operators. Each LNS version starts with two different initial solutions. At each iteration, LNS ejects a given number of customers using its removal operator. Then, LNS inserts a set of customers using its repair operator. After an experimental study presented in [15], we choose the random removal and the regret heuristic as the pair of removal/insertion operators used in LNS.

Our LNS uses the Simulated Annealing principle as acceptance criteria. Hence, a solution is accepted if it is better than the current solutions or, if the metropolis rule is satisfied. In other words, bad solutions are accepted with the probability given by $\exp^{\frac{f(s)-f(s')}{T}}$, where $f(.)$ stands for the objective function, s' and s are the new and the current solutions respectively, and $T > 0$ denotes the current temperature. At each iteration, the current temperature T is decreased using the cooling rate $c \in]0,1[$ according to the expression $T = T \times c$. For more information regarding the selection of the removal/insertion couple, we refer the reader to [15]. The general schema is depicted in Fig. 3.

3.6   Adaptive Large Neighborhood Search Framework

The ALNS framework was first proposed by [20] and can be described as a Large Neighbourhood Search algorithm with an adaptive layer, where a set of destroy/repair operators compete to modify the current solution in each iteration of the algorithm. When compared to LNS, ALNS makes more drastic changes to the current solution by exploring an even larger search space. Recent Methods in the literature based on ALNS present a noticeable success for several variants of vehicle routing problem [2,4,10]. The success of the method comes mostly from its capacity to search a more complicated neighbourhood in comparison to other methods applied to vehicle routing problem and its ability to dynamically controls the probability of selecting the destroy/repair operators according to their performance history in the search process [19].

Our ALNS uses three destroy operators and three repair operators described in Sects. 3.2 and 3.3. The general ALNS scheme is similar to the LNS scheme (see Fig. 3) except that ALNS uses an adaptive mechanism to select the operators that are most suitable for the search space.

In practice, the ALNS starts with identical π_i scores for each operator i and each time a destroy/repair operation is performed, the scores of the operators concerned are increased. A score can be increased by:

– σ_1 if a new, better global solution is found,
– σ_2 if a solution not yet accepted is found and this solution has a better quality than the current solution,
– σ_3 if the solution found is worse than the current solution and it is still accepted knowing it's never been accepted before.

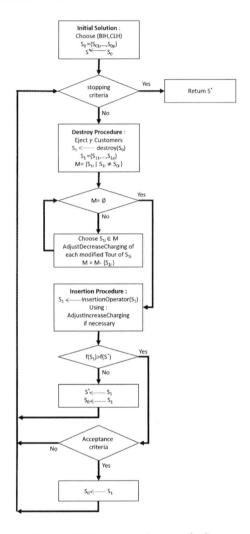

Fig. 3. LNS scheme (source: [15]).

Each time a certain number of predetermined iterations is reached, we'll say that ALNS has moved on to a new segment, named Seg. The size of this segment will be determined in the experimentation section. Therefore, the method computes the weight $w_{i,j+1}$ of each operator i for the new segment $j+1$ according to the weight $w_{i,j}$ of this operator during the previous segment j. We use the formula: $w_{i,j+1} = w_{i,j} \times (1 - \lambda) + \lambda \times \frac{\pi_i}{\theta_i}$, where λ is a parameter controlling the reaction rate of the algorithm at the score changes, π_i and θ_i represent, respectively, the score and the number of times that operator i was used during the segment j. The selection of operators is then performed, at each iteration of the segment $j+1$, by using the Roulette-Wheel Selection principle. Thus, an operator i, who has had a high weight during the previous segment, will have a better chance of being selected. More precisely, the chance of selecting such an operator is $(w_i / \sum_{a \in O} w_a)$, where O is the set of all operators of the same type (removal or repair).

We use as acceptance criteria for ALNS the Metropolis rule (simulated annealing) which is the same one used for LNS.

4   Computational Tests

Our methods are implemented using C++. All experimentations are carried out on an Intel Core (TM) i7- 5600U CPU, 2.60 GHz processor, with 8 GB RAM memory. We use instances with 100 clients proposed in [16].

Based on the study performed in [15], we set the parameters of our LNS as follows:

- Number of iterations = 1000
- Number of clients to reject $\gamma = 20$.
- Insertion strategy for stations charging $\beta = All$
- Insertion strategy for customers $\alpha = All$
- Acceptance criteria: the initial temperature is defined in a way that at first iterations of LNS, a solution 20% worse than the current one will be accepted with probability of 0.5, and the cooling rate is initialised to 0.9995.

The parameters described above for LNS are also used for our ALNS. Preliminary tests have confirmed the validity of these parameters for ALNS framework. To set the remaining parameters $(Seg, \sigma_1, \sigma_2, \sigma_3)$, we performed a series of tests on 20 instances and 4 combinations of parameters $(Seg, \sigma_1, \sigma_2, \sigma_3)$. These combinations are mentioned in the columns of Table 1. We save the best solution, named *BestSol* found on all tests. In rows 1 of Table 1, we computed the number of times that ALNS with the chosen combination finds the best solution on all the tests. Line2 (respectively line 3) shows the average deviation (respectively the worst deviation) from *BestSol*. The results of Table 1 show that none of the ALNS variants dominates all the others, and that the combination of column 2 seems to be a good compromise since this variant never achieves bad results on the three criteria, and succeeds in obtaining good results on the first two criteria. We then decided to retain the values $(50, 33, 10, 15)$ for the combination $(Seg, \sigma_1, \sigma_2, \sigma_3)$.

In Table 2, we compare our LNS variant that uses the best pairs of operators (random removal and regret insertion) with ALNS. Column 2 (respectively column 5) gives the initial solution obtained with BIH (respectively CLH). Column 3 (respectively column 6) gives the deviation of LNS solution in relation to the BIH initial solution (respectively CLH initial solution). Column 4 (respectively column 7) gives the deviation of ALNS solution in relation to the BIH initial solution (respectively CLH initial solution). For a given instance, we note, in bold, the best solution found (Best Cost). The results indicate that LNS can improve the initial solution on average by up to 34%. LNS succeeds in improving the initial solution in all cases except in the instance 3. The best solution is obtained in all cases by LNS. LNS using BIH as initial solution succeeds in attaining the best solution in 6 cases while the LNS using CLH reaches the best solution in 2 cases. The ALNS can improve the initial solution on average by up to 29%. It reaches the best solution only once. We can conclude that LNS remains very competitive compared to ALNS. In our case, an adaptive approach does not bring any improvement because the problem is very difficult (the number of feasible solutions can be limited), and therefore LNS for which the choice of operators has been optimized could already visit enough neighbourhood. In fact, including several repair and destroy operators in classical framework of ALNS does not allow the method to have a larger neighbourhood or larger search space. This remark can be confirmed by examining Fig. 4. Indeed, the number of new best solutions found during the iterations of ALNS remains very small and the number of new worst solutions remains very significant. We mention that the average computation time for the two methods is almost similar (2.9 h on average for LNS and 2.4 h on average for ALNS). We can conclude that a classic version of ALNS like the one proposed in [20] is not competitive with LNS. ALNS needs other ingredients to enhance its results, such as the integration of a diversification mechanism and local search techniques.

Table 1. $(Seg, \sigma_1, \sigma_2, \sigma_3)$ setting parametters for ALNS.

	(50,33,15,10)	(50,33,10,15)	(100,33,15,10)	(100,33,10,15)
NBestSol	7	6	4	3
Average Gap	2,42%	1,97%	2,05%	2,38%
Worst Gap	6,45%	6,27%	5,41%	6,17%

To understand the convergence of the LNS and ALNS methods during iterations. We computed at each 250 iterations, the number of new best solutions found. More precisely, for each interval $]0, 250]$, $]250, 500]$, $]500, 750]$, $]750, 1000]$, we compute the average of the number of times that the best solution was reached on all instances. The values are not cumulative, i.e. if the best solution is reached in one interval, we do not consider this solution in the other intervals. We only consider the new best solutions that will be obtained in the concerned interval. The results are displayed in the Table 3.

The results reveal that LNS and ALNS have almost the same behaviour. About half of the best solutions are found between 0 and 500. We can notice that the percentage of the best solutions found is not negligible between 750 and 1000 iterations especially for LNS. Following these results, we decide to examine whether the percentage of improvement is significant enough at last iterations of LNS/ALNS. In Fig. 5, we give for LNS using BIH initial solution, LNS using CLH initial solution, and ALNS, the evolution curve of the average gap in relation to the initial solution on all instances using the results obtained

Table 2. Computational results of LNS and ALNS.

	BIH			CLH			
	S_0 Cost	LNS Gap_{S0}	ALNS Gap_{S0}	S_0 Cost	LNS Gap_{S0}	ALNS Gap_{S0}	Best Cost
Inst1	1600,28	8%	0%	2050,72	**29%**	22%	1466,23
Inst2	1543,80	0%	0%	2062,29	**25%**	**25%**	1539,83
Inst3	**1596,50**	0%	0%	2260,59	27%	22%	1596,50
Inst4	3798,70	**57%**	54%	4543,74	63%	61%	1642,74
Inst5	3247,57	**43%**	36%	3476,44	43%	38%	1842,20
Inst6	2751,84	**31%**	20%	2818,55	29%	24%	1906,10
Inst7	2491,52	**27%**	20%	2663,88	29%	24%	1816,03
Inst8	2826,77	**28%**	19%	3219,18	33%	31%	2042,13
Inst9	2836,07	**29%**	23%	2825,21	28%	20%	2008,86
Avg		25%	19%		34%	29%	

Table 3. Percentage of the number of BestSol each 250 iterations.

	Iterations				
	0	]0,250]	]250,500]	]500,750]	]750,1000]
BIH+LNS	22%	11%	11%	11%	44%
BIH+ALNS	33%	11%	22%	0%	33%
CLH+LNS	0%	0%	33%	22%	44%
CLH+ALNS	0%	33%	33%	22%	11%

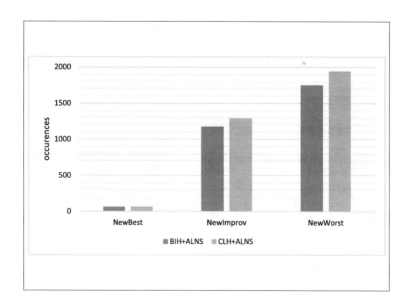

Fig. 4. Comparison of the solutions quality found in the ALNS process.

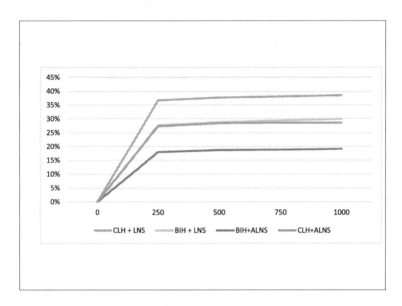

Fig. 5. Evolution of the average GAP in LNS/ALNS.

at the end of each iteration 250, 500, 750 and 1000. We note that the convergence is almost achieved at iteration 250 and that after 250 iterations the improvement is very small therefore running the LNS or ALNS for 250 iterations is a good compromise to reduce the execution time.

5   Conclusion

In this paper, we proposed two metaheuristics based on large neighbourhood search to deal with a Periodic Electric Vehicle routing problem (PEVRP). The PEVRP consists in optimizing the routing and charging of a fixed set of vehicles over a multi-period horizon. We proposed three destroy and three repair operators and specific procedures to address energy constraints. These specific procedures are applied each time a re- pair/destroy operator is used. We tested and compared two metaheuristics approaches that have been proven to be effective on vehicle routing problems. The first one uses a single operator pair and it is known as LNS, so we optimized the choice of the pair of operators, to be integrated in LNS, through a preliminary study conducted in [15]. The second one uses all operators in a classic ALNS framework, in which all operators compete to modify the current solution in each iteration of the ALNS algorithm using an adaptive mechanism. The comparison of the two metheuristics revealed that LNS remains competitive compared to ALNS. Indeed, ALNS does not allow covering a wider neighbourhood since the problem is very constraining (mainly because of energy constraints). These results are quite interesting and show that a simple method like LNS could be interesting on complex problems. We are currently pursuing this study to determine whether a more improved version of ALNS would enhance ALNS's results and allow it to be competitive with LNS.

References

1. Archetti, C., Fernández, E., Huerta-Muñoz, D.L.: A two-phase solution algorithm for the flexible periodic vehicle routing problem. Comput. Oper. Res. **99**, 27–37 (2018)
2. Azi, N., Gendreau, M., Potvin, J.Y.: An adaptive large neighborhood search for a vehicle routing problem with multiple routes. Comput. Oper. Res. **41**, 167–173 (2014)
3. Baldacci, R., Bartolini, E., Mingozzi, A., Valletta, A.: An exact algorithm for the period routing problem. Oper. Res. **59**(1), 228–241 (2011)
4. Chentli, H., Ouafi, R., Cherif-Khettaf, R.W.: A selective adaptive large neighborhood search heuristic for the profitable tour problem with simultaneous pickup and delivery services. RAIRO-Oper. Res. **52**(4–5), 1295–1328 (2018)
5. Christofides, B.J.: The period routing problem. Networks **14**, 237–256 (1984)
6. Dayarian, I., Crainic, T., Gendreau, M., Rei, W.: An adaptive large-neighborhood search heuristic for a multi-period vehicle routing problem. Transp. Res. Part E **95**, 95–123 (2016)
7. Erdelić, T., Carić, T.: A survey on the electric vehicle routing problem: variants and solution approaches. J. Adv. Transp. **2019** (2019)
8. Felipe, M., Ortuno, T., Righini, G., Tirado, G.: A heuristic approach for the green vehicle routing problem with multiple technologies and partial recharges. Transp. Res. Part E: Logistics Transp. Rev. **71**, 111–128 (2014)
9. Francis, P.M., Smilowitz, K.R., Tzur, M.: The period vehicle routing problem and its extensions. In: Golden, B., Raghavan, S., Wasil, E. (eds.) The Vehicle Routing Problem: Latest Advances and New Challenges. Operations Research/Computer Science Interfaces, vol. 43, pp. 73–102. Springer, Boston (2008). https://doi.org/10.1007/978-0-387-77778-8_4
10. Ghilas, V., Demir, E., Woensel, T.V.: An adaptive large neighborhood search heuristic for the pickup and delivery problem with time windows and scheduled lines. Comput. Oper. Res. **72**, 12–30 (2016)
11. Goeke, D., Schneider, M.: Routing a mixed fleet of electric and conventional vehicles. Eur. J. Oper. Res. **245**(1), 81–99 (2015)
12. Goeke, D.: Granular tabu search for the pickup and delivery problem with time windows and electric vehicles. Eur. J. Oper. Res. **278**, 821–836 (2019)
13. Hiermann, G., Puchinger, J., Ropke, S., Hartl, R.: The electric fleet size and mix vehicle routing problem with time windows and recharging stations. Eur. J. Oper. Res. **252**(3), 995–1018 (2016)
14. Jie, W., Yang, J., Zhang, M., Huang, Y.: The two-echelon capacitated electric vehicle routing problem with battery swapping stations: formulation and efficient methodology. Eur. J. Oper. Res. **272**(3), 879–904 (2019)
15. Kouider, T.O., Ramdane Cherif-Khettaf, W., Oulamara, A.: Large neighborhood search for periodic electric vehicle routing problem. In: Proceedings of the 8th International Conference on Operations Research and Enterprise Systems - Volume 1: ICORES, pp. 169–178. INSTICC, SciTePress (2019)

16. Kouider, T.O., Ramdane-Cherif-Khettaf, W., Oulamara, A.: Constructive heuristics for periodic electric vehicle routing problem. In: Proceedings of the 7th International Conference on Operations Research and Enterprise Systems - Volume 1: ICORES, pp. 264–271. INSTICC, SciTePress (2018)
17. Mancini, S.: A real-life multi depot multi period vehicle routing problem with a heterogeneous fleet: formulation and adaptive large neighborhood search based matheuristic. Transp. Res. Part C **70**, 100–112 (2016)
18. Pelletier, S., Jabali, O., Laporte, G.: Goods distribution with electric vehicles: review and research perspectives, working paper, CIRRELT (2014)
19. Pisinger, D., Røpke, S.: Large Neighborhood Search, 2 edn., pp. 399–420. Springer, Heidelberg (2010)
20. Ropke, S., Pisinger, D.: An adaptive large neighborhood search heuristic for the pickup and delivery problem with time windows. Transp. Sci. **40**, 455–472 (2006)
21. Sassi, O., Ramdane-Cherif-Khettaf, W., Oulamara, A.: Iterated tabu search for the mix fleet vehicle routing problem with heterogenous electric vehicles. Adv. Intell. Syst. Comput. **359**, 57–68 (2015)
22. Sassi, O., Cherif-Khettaf, W.R., Oulamara, A.: Multi-start Iterated Local Search for the Mixed Fleet Vehicle Routing Problem with Heterogenous Electric Vehicles. In: Ochoa, G., Chicano, F. (eds.) EvoCOP 2015. LNCS, vol. 9026, pp. 138–149. Springer, Cham (2015). https://doi.org/10.1007/978-3-319-16468-7_12
23. Schiffer, M., Walther, G.: The electric location routing problem with time windows and partial recharging. Eur. J. Oper. Res. **260**, 995–1013 (2017)
24. Schiffer, M., Walther, G.: Strategic planning of electric logistics fleet networks: a robust location-routing approach. Omega **80**, 31–42 (2018)
25. Schneider, M., Stenger, A., Goeke, D.: The electric vehicle routing problem with time windows and recharging stations. Transp. Sci. **75**, 500–520 (2014)

A Matheuristic for the Design and Management of Multi-energy Systems

A. Bartolini[1], G. Comodi[1], F. Marinelli[2], A. Pizzuti[2(✉)], and R. Rosetti[2]

[1] DIISM - Dipartimento di Ingegneria Industriale e Scienze Matematiche,
Universitá Politecnica delle Marche, 60131 Ancona, Via Brecce Bianche, Italy
`a.bartolini@pm.univpm.it, g.comodi@staff.univpm.it`
[2] DII - Dipartimento di Ingegneria dell'Informazione,
Universitá Politecnica delle Marche, 60131 Ancona, Via Brecce Bianche, Italy
`{fabrizio.marinelli,r.rosetti}@univpm.it, a.pizzuti@pm.univpm.it`

Abstract. New technologies and emerging challenges are drastically changing how the energy needs of our society have to be met. By consequence, energy models have to adapt by taking into account such new aspects while aiding in decision making processes of the design of energy systems. In this work the problem of the design and operation of a multi-energy system is tackled by means of a mixed integer linear programming (MILP) formulation. Given the large size of the problem to be solved, a matheuristic approach based on constraint relaxations and variable fixing is proposed in order to not restrict the applicability to small cases. Two variable fixing policies are presented and performance analysis comparison on them has been done. Tests have been performed on small and realistic instances and results show the correctness of the approach and the quality of the heuristic proposed in term of solution quality and computational time.

Keywords: Multi-energy system management · District design · MILP · Matheuristic

1 Introduction

The decarbonization of our society is a challenge which is becoming increasingly urgent due to the effects of climate change, for this reason many countries and organizations will or have already started to commit in addressing such challenge by establishing policies aimed at the reduction of greenhouse gases emissions, e.g. [1,2]. Tthe reduction of such carbon footprint can be achieved following several paths, with one of the most common being the increase of renewable energy technologies to meet the diverse needs of final users. The capacity of renewable energy sources is thus increasing worldwide, especially regarding wind and solar electricity generation systems, also thanks to their investment costs which are expected to drop following both research and development and improved economies of scale as confirmed by several studies as [3] and [4]. But both of

© Springer Nature Switzerland AG 2020
G. H. Parlier et al. (Eds.): ICORES 2019, CCIS 1162, pp. 171–188, 2020.
https://doi.org/10.1007/978-3-030-37584-3_9

these technologies come with the drawback of an intermittent and non controllable electricity output, which with high capacities poses significant challenges on the current electricity production and distribution infrastructures as highlighted in [5]. Such unpredictability can be addressed following several paths, the more common has historically been to rely on peaker plants (fast ramping natural-gas powered electricity generation systems) or large electricity storage such as pumped hydro or, more recently, large centralized battery storage systems [6]. But both approaches come with drawbacks: the possibility of relying on a pumped hydro storage system depends on local conditions, and peaker plants (other thank being a source of carbon emissions themselves) are usually expensive to run. However a more recent approach lies in addressing such challenge thus lies in decentralizing the energy distribution system by means of microgrids and distributed multi energy systems [7]. Microgrids have been an established solution to assess the needs of rural off-grid communities, but the advantages they bring are gaining interest from researchers also regarding already urbanized contexts as in [8] and [9].

Microgrids are a concept strictly related to electricity management, but with a distributed multi energy systems (MES) approach the concept is widened to simultaneously consider multiple energy vectors (i.e., carriers of energy as electricity) [10], with more opportunities for an efficient use of primary energy resources that can be exploited thanks to such interactions [11,12]. In order to achieve the best results from such approach all of the potential needs (in terms of demanded commodities) of the users of a MES have to be considered: such as space heating/cooling, hot water, electricity etc. Modern energy systems models have then to take into consideration such interactions, while also correctly assessing the potential presence of high shares of non controllable renewable electricity generation technologies as highlighted in [13].

In [14] a mixed integer linear programming (MILP) formulation was proposed to model the optimal design and management of MES, by simultaneously evaluating the investment in new technologies and the provision of energy commodities by external energy distribution infrastructures in order to meet the energy demands of the users while satisfying operational constraints, such as the time varying dispatch of the available systems. Based on the proposed MILP, a matheuristic approach was designed to handle instances of practical interest. The matheuristic is based on relaxing integrality constraints and fixing subsets of variables that are chosen accordingly to a commodity-based decomposition strategy. Preliminary experiments were performed to analyze the performance of the heuristic algorithm in terms of solutions quality and computational effort.

To the best of our knowledge, other approaches based on fixing integer variables and relaxing constraints has only been applied by [16] for energy systems related problems, where the optimal design for a rural electrical microgrid is evaluated using different heuristic approaches, among which a fix-and-relax strategy. In the last few years the optimal design and operation management of distributed MES has been addressed by means of several methods [17]. The problem has been formalized with both single or multi-objective functions, and solved by techniques that can be mainly classified into single monolithic and decomposition approaches. With respect to the former, [18] and [19] exploit MILP formulations to compute the optimal design of a small district needing electricity and heat; the same is done in [20] where the demand of cooling power is also considered. For the latter, decomposition approaches rely on keeping the planning phase, that is the decision to purchase and deploy, separated by the operational phase. In [21] a mixed integer non linear programming (MINLP) formulation is presented and solved by a two phases approach, where the first is handled by means of two evolutionary algorithms and a discrete variable relaxation technique, while the second is solved by exploiting a linearized MILP. In [22] the design of an hydrogen based microgrid is performed by using a genetic algorithm to determine the size of the systems, while the operational variables are set by solving a MILP. A similar technique is described in [23], where the same framework is applied for a multi-objective variant of the problem.

In this paper, we extend the contents previously presented in [14] by proposing an alternative fixing strategy for the matheuristic that relies on partitioning the time horizon into sub-intervals, fixing the values of binary variables that model the operation of technologies within such sub-intervals. Moreover, a back-track procedure is integrated in the matheuristic to deal with unfeasible states. Finally, we enlarge the test-bed by considering a wider set of randomly generated instances and additional realistic instances derived from two case studies. Computational experiments are discussed to compare the quality of the two fix-and-relax strategies.

The rest of the paper is structured as follows: in Sect. 2 the description of the used energy hub and technology models is given, in Sect. 3 the problem is formalized and the mathematical model is described; in Sect. 4 the matheuristic algorithm is presented and the two fixing strategies are depicted in 4.1 and 4.2, respectively; computational results based on both randomly generated instances and the two case studies are given in Sect. 5. Finally in Sect. 6 conclusions and future perspectives are discussed.

2   Energy Hub and Technology Modelling

In this section the concept of energy hub and its implementation for the proposed case studies in Sect. 5.2 are described. Furthermore, the technologies that are considered within the analyses and their general functioning principles are also described.

2.1   The Energy Hub

A comprehensive description of the concept of energy hub has been given in [24]. Energy hubs can be thought as integrated systems within which different energy vectors are converted by means of technologies in order to meet the needs of a user located within the same hub, or else way sent to a different energy hub located elsewhere. But an energy hub cannot produce any energy vector by itself, thus energy is provided from outside by means of existing infrastructures (electricity/natural gas distribution networks) or natural resources (wind, solar radiation, biomass etc.). Then, provided that the appropriate technology is either already present or purchased, such energy vectors are converted to meet the needs of the users. The total costs sustained for the proper functioning of the hub can then be divided in investments needed to purchase new systems and operational costs needed for their functioning, such as for example the purchasing of natural gas or electricity from the external distribution networks. A realistic test case is then modeled as an energy hub in order to test the proposed algorithm, with the proposed energy hub structure shown in Fig. 1. The test case is a community consisting in buildings of the tertiary sector, which in order to properly function need space heating during the winter, space cooling during the summer, and electricity throughout the whole year. It's considered to be a community within a developed context, thus with access to reliable electricity and natural gas distribution infrastructures.

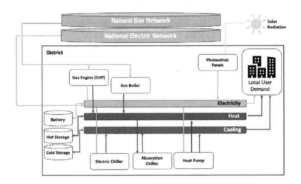

Fig. 1. Proposed energy hub model [14].

2.2   Technologies Modeling

Within this study a set of already mature and commercially available technologies have been considered for potential deployment within the energy hub, and such technologies include both conversion technologies (conversion between energy vectors) and storage ones (storage of an energy vector for later use). These can be listed as:

- **Natural Gas Engine** (CHP): operates in combined heat and power (CHP) mode, thus producing electricity and simultaneously recovering part of the waste heat, using natural gas as fuel;
- **Natural Gas Boiler** (GB): production of heat from natural gas;
- **Electric Chiller** (EC): conversion of electricity into cooling energy;
- **Absorption Chiller** (AC): conversion of heat into cooling energy;

- **Heat Pumps** (*HP*): allow the conversion of electricity into (alternatively) cooling energy or heat, based on the required output;
- **Photovoltaic Panels** (*PV*): production of electricity from solar radiation;
- **Batteries** (*EES*): storage of electricity;
- **Hot Thermal Storage** (*HTES*): storage of heat;
- **Cold Thermal Storage** (*CTES*): storage of cooling energy.

A small database of purchasable devices is available for each of such technologies, distinguished by size. Then for each particular size, parameters representing costs and technical performance are specified. The performance of the conversion devices is described by an energy conversion efficiency for each available size, whereas the relevant parameters for the storage devices are the efficiency for both the charge and the discharge phases. Besides, conversion devices are enforced to work with an output above a fixed threshold value for a certain timespan, this to represent a lower bound of partialisation (i.e., the minimum output power) for the technology. The only exception concerns the photovoltaic panels, whose produced electric energy is directly dependent on the total panel surface and the solar exposition, which is a function of the timestep. Hence, in absence of solar radiation the output of the photovoltaic panels is zero.

Regarding costs, each model of technology is characterised by its own purchase cost related to its size and performance. Such cost has to be sustained in order to make use of the technology for a representative lifetime in years (which is also provided as technical input parameter), after which the purchase cost has to be re-sustained. Moreover, additional costs related to the maintenance operations are considered. In this way, all of the costs related to purchasing and operating phases of a certain technology are taken into account. Finally, further costs sustained within the energy hub are given by the possible withdrawal of energy from the two national infrastructures. The withdrawal is priced with a fixed cost per kWh of purchased energy, different for natural gas and electricity, and can be done in unlimited quantities of energy withtin a single timestep.

3   District Energy Design and Scheduling: Problem Definition and Mathematical Formulation

The energy design of the district and the scheduling of resources can be described as follows: an energy district has to be constructed and managed in order to satisfy the demand of different energy vectors (commodities) by selecting and controlling an equipment of technologies. Let K be the set of commodities and c_k and h_k the fixed unit purchasing and storing price respectively. Set Q represents the set of technology models that can be deployed in the district.

The set $S \subseteq Q$ is defined as the subset of all the storage systems and $S^k \subseteq S$ is the set storage models that can manage the commodity k. Parameter v_i is defined for each $i \in Q$, stating the costs for buying, installing and maintaining the i-th appliance type. This cost is normalized along the considered period: indeed, the approximated profile of the demand of each commodity is reported for a discrete time horizon $T = \{1, \ldots, \tilde{T}\}$ with intervals of one hour; in particular d_t^k is the demand of the commodity $k \in K$ at time $t \in T$.

Therefore, the matrix $\mathbf{v}$ is normalized in this way: given the purchasing and installing cost (IC), maintenance cost per year (MCY) and estimated lifetime (expressed in year and hours respectively as ELY and ELH), each element of $\mathbf{v}$ is defined by:

$$v_i = \frac{IC_i + MCY_i \times ELY_i}{ELH_i} \times \tilde{T}$$

In this way, even if usually fixed costs are all sustained in the same moment, the definition of v_i includes only the portions of hourly costs related to the considered time horizon T. Hence, the ratio between fixed and variables costs (e.g. c_k) is contained, while the objective function weights the total cost associated to the life cycle of each device.

Decisions have to be performed such that the total costs sustained by the energy district within a certain time horizon are minimized, both in the design and scheduling phases. The time horizon is cyclically repeated to describe the optimal dynamic of a fully operational energy district, satisfying requests that (hopefully) follow periodical patterns.

Technologies are very different one from the other as they can provide different types of commodities using others. For example a Gas Engine uses gas and can supply heat or electricity, whereas an Electric Chiller needs electricity to work and supply cool. The parameter ω_i^{hk} describes the conversion technology $\forall i \in Q \setminus S$, expressing the unitary conversion multiplier to convert commodity h into commodity k by means of appliance i. Devices related to green energy, such as solar panels or wind farms, have a base production that vary according to weather conditions. Thus, parameter b_{it}^k states the base production of equipment $i \in Q \setminus S$ at time t of commodity $k \in K$. The amount of output k produced by the conversion device i at each time t is ranged within the maximum rated power U_i^k and the lower bound L_i^k.

Similarly, storage technology $i \in S^k$ has a capacity C_i^k whose reuse is regulate by an efficiency parameter ϕ_k. In fact, the quantity of commodity k collected is subjected to a dispersion factor (or, as previously said, efficiency parameter).

Non-negative variables $\mathbf{r}$ and $\mathbf{p}$ are defined to model the flow of commodities transformed by conversion devices. Variable $r_{it}^{hk} \in \mathbb{R}^+$ expresses the amount of commodity h converted into commodity k by device $i \in Q \setminus S$ at each time instant t. On the other hand, variable $p_{it}^k \in \mathbb{R}^+$ is the summation of the eventual base productions b_{it}^k (e.g. solar energy) and the quantity of commodity k converted from other commodities h represented by variable r_{it}^{hk} (properly scaled by parameter ω_i^{hk}). Furthermore, for each $k \in K$ and $t \in T$ variables l_t^k and f_t^k are defined to describe for time slot t the total amount of commodity k stored within the district and the quantity of k acquired by the exogenous supplier respectively.

Binary variables $y_i \ \forall i \in Q$ are introduced to model the selection of devices, i.e. $y_i = 1$ if and only if the district is provided with equipment i since the beginning of T. Since each technology has an estimated lifetime much larger than $\tilde{T}$, no considerations are necessary for the renew of technologies while mmaintenance is considered in the buying cost. Devices operativity are finally described by variables $z_{it} \in \{0,1\}$ defined $\forall i \in Q \backslash S$ and $\forall t \in T$ that get value 1 if device i is working during timeslot t.

The MILP formulation reads as follow [14]:

$$\min \sum_{t \in T} \sum_{k \in K} (c_k f_t^k + h_k l_t^k) + \sum_{i \in Q} v_i y_i \tag{1}$$

$$d_t^k + l_t^k + \sum_{i \in Q \backslash S} \sum_{\substack{h \in K: \\ h \neq k}} r_{it}^{kh} =$$

$$\phi_k l_{t-1}^k + \sum_{i \in Q \backslash S} p_{it}^k + f_t^k \quad \forall k \in K, \forall t \in T \tag{2}$$

$$p_{it}^k = \sum_{\substack{h \in K: \\ h \neq k}} \omega_i^{hk} r_{it}^{hk} + b_{it}^k y_i \quad \forall i \in Q \backslash S, \forall k \in K, \forall t \in T \tag{3}$$

$$L_i^k z_{it} \leq p_{it}^k \leq U_i^k z_{it} \quad \forall i \in Q \backslash S, \forall k \in K, \forall t \in T \tag{4}$$

$$z_{it} \leq y_i \quad \forall i \in Q \backslash S, \forall t \in T \tag{5}$$

$$l_t^k \leq \sum_{i \in S^k} y_i C_i^k \quad \forall k \in K, t \in T \tag{6}$$

$$l_0^k = l_T^k \quad \forall k \in K \tag{7}$$

$$l_t^k, f_t^k \geq 0 \quad \forall k \in K, \forall t \in T \tag{8}$$

$$p_{it}^k \geq 0 \quad \forall i \in Q \backslash S, \forall k \in K, \forall t \in T \tag{9}$$

$$r_{it}^{hk} \geq 0 \quad \forall i \in Q \backslash S, \forall h, k \in K : h \neq k, \forall t \in T \tag{10}$$

$$y_i \in \{0,1\} \quad \forall i \in Q \tag{11}$$

$$z_{it} \in \{0,1\} \quad \forall i \in Q \backslash S, \forall t \in T \tag{12}$$

The objective function (1) aims to minimize the total cost given by the cost related to deployed technologies, the cost of acquiring external commodities from public suppliers and the expense for storing surpluses for further use. Constraints (2) are stock-balancing constraints and state that the commodity k demanded, stored and required by devices is equal to the summation among the previously amount of commodity k stored at time $t-1$, the total one produced by all deployed technologies and the quantity acquired from external networks. The set of constraints (3) defines the variables $\mathbf{p}$, for each conversion technology $i \in Q \backslash S$, as the summation between the required commodity h necessary to produce commodity k and a base production without consumption. This is due to the feature of green technologies, such as photovoltaic and wind farm, to produce commodities without consuming other resources. The bounds of variables $\mathbf{p}$ are given by the set of constraints (4): if binary variable z_{it} is equal to 1, the corresponding variable p_{it}^k is limited between values L_i^k and U_i^k. The link among variables $\mathbf{z}$ and $\mathbf{y}$ is coherently modelled by the set of constraints (5). Constraints (6) limit the stored quantity of commodity $k \in K$ at any time $t \in T$

to the total sum of capacities of the storage systems in S^k installed. In order to translate the cyclicality of the selected time horizon, the quantities of commodities stored in the first timestep $(t = 1)$ are balanced in (2) with the quantities stored in the last timestep $(t = \tilde{T})$ by means of equality (7). It is worth noting that whenever an appliance i is not able to convert a certain commodity h into k, the corresponding parameter ω_i^{hk} is null. In all such cases, $\forall t \in T$ variables r_{it}^{hk} are fixed to zero as a preprocessing phase. Moreover, $\forall t \in T$ variables f_t^k are set equal to zero for any commodity $k \in K$ that is not purchasable by external suppliers.

4   Heuristic Algorithm: Idea and Approaches

The mathematical formulation proposed in Sect. 3 cannot be exploited to solve real instances in a reasonable amount of time by means of standard solvers as computational experiments showed. This can be (partially) related to the size of the mathematical model, as the number of constraints is $\mathcal{O}(|Q||K||T|)$, the number of binary and continuous variables are $\mathcal{O}(|Q||T|)$ and $\mathcal{O}(|Q||K|^2|T|)$ respectively. As example, lets consider an instance with 4 commodities, 15 conversion devices and 5 storage systems, with an horizon of 100 time steps. Applying the mathematical model to this instance, that is far from a realistic case, the formulation already counts 24800 continuous variables, 1520 binary variables and more than 20000 constraints. Therefore, a matheuristic algorithm to approach instances that are meaningful in real context have been designed. The matheuristic used to solve the proposed MILP is based on relaxing integrality constraints and fixing subsets of variables, similarly to the *Fix & Relax* scheme formalized in [15].

The idea is to solve the MILP starting with its linear relaxation and build a feasible solution by restoring the integrality conditions only for a subset of the original integer variables.

This "partial" linear relaxation is solved and the integer variables are fixed to their values in the resulting solution in order to reduce the number of variables and shrunk the space of research. Iterating this process, with the necessary technical devices, leads to a feasible integer solution.

A common challenge for this kind of approach is to find an equilibrium between the algorithm efficiency and quality of the solution. In the approach proposed, a small subset of variables with integrality restored lead to a faster integer solution, but very small set can deteriorate solution quality rapidly. Moreover, a robust strategy is one that can avoid the convergence to an unfeasible solution, or is able to efficiently backtrack whenever this occurs.

The *Fix & Relax* scheme governs the matheuristic approach adopted for the resolution of the energy district design and management problem A customization of this scheme can be obtained by specifying the way of selecting and fixing of the subset of binary variables. For this purpose, two different schemes have been thought up: in the former, binary variables are obtained by separating the corresponding technologies for their capacity of converting or storing commodities; for the latter instead, the subsets of binary variables are identified after sampling subsets of timesteps from the time horizon. In the following, each strategy will be explained in detail.

4.1   Commodity-Based Scheme $\mathcal{H}_K$

The idea behind the framework $\mathcal{H}_K$ relies on the fact that each technology rarely produce (store) more than one commodity, thus creating a natural separation between them. This division can then be translated into a clustering criterion for grouping the binary variables $\mathbf{y}$. Moreover, as detailed below, the fixing policy is defined to allow a certain grade of flexibility to the matheuristic and to reduce the possibility of reaching unfeasible states.

The scheme can be summarized as follow. Let us consider the mathematical formulation proposed in Sect. 3 as $\mathcal{M}$. Let V_y and V_z be the sets of variables in $\mathcal{M}$ for which the binary condition holds, related respectively to $\mathbf{y}$ and $\mathbf{z}$. In a similar way, let us define V_y^r and V_z^r as the subsets of variables whose integrality is relaxed in $\mathcal{M}$. Finally, let K_r be an auxiliary set initialized as K and collecting all the commodities. As first step, the algorithm considers the linear relaxation of $\mathcal{M}$ with $V_y = V_z = \emptyset$. Considering the demands and costs for acquiring external resources, a priority policy for commodity selection is defined among the commodities $k \in K$. In details, at each iteration the commodity $\bar{k} \in K_r$ is selected accordingly to the following rule:

$$\bar{k} = \arg\max_{k \in K_r}\{d_{max}^k / c_k\} \tag{13}$$

where $d_{max}^k = \max_{t \in T}\{d_t^k\}$. In other words, the commodity chosen has the largest ratio between its own demand peak and the unity of cost for the external supply among all the commodities $k \in K_r$. All the variables $y_i \in V_y^r$ modeling appliances able to convert (store) the commodity $\bar{k}$ have the integrality constraints restored and commodity $\bar{k}$ is then removed from K_r. In other words, the binary condition for variables y_i is restored for all the conversion technologies $i \in Q \setminus S$ with $\omega_i^{h\bar{k}} > 0, \forall h \in K : h \neq \bar{k}$. Set V_y^r and V_y are updated according to the integrality restored, that is $V_y^r = V_y^r \setminus \{y_i\}$ and $V_y = V_y \cup \{y_i\}$. This criteria involves the storage systems able to store the commodity chosen, that is, the storage systems $i \in S$ with positive $C_i^{\bar{k}}$ have restored the integrality condition on the variables y_i associated.

Formulation $\mathcal{M}$ is solved and a fixing procedure is invoked to establish the value of the variables just put in V_y. In fact, for each variable y_i in V_y, its value is fixed to one if and only if $y_i = 1$ in the resulting solution. Variables y_i, with value equal to zero in the resolution are fixed to zero only for those who are related to conversion devices $i \in Q \setminus S$. By excluding the variables related to the storage devices, the flexibility of the current solution is partially preserved. In fact, if in a later iteration related to $h \neq \bar{k}$ the requests $r_{it}^{\bar{k}h}$ increase due to deployment of different appliances, then storage devices of $\bar{k}$ are still valid options to avoid the increment of $f_t^{\bar{k}}$. In order to enhance the efficiency, a trade-off is used to fix the variables not in V_y. In fact, each y_i in V_y^r such that $y_i > 1 - \epsilon$ (with ϵ reasonably small) is fixed to one and moved from V_y^r to V_y. The same applies to variables vector $\mathbf{z}$. For each y_i currently fixed to 1, all the corresponding z_{it} in V_z^r are fixed to one if $z_{it} > 1 - \epsilon$, and moved from V_z^r to V_z. This process is iterated until K_r is empty, i.e. after $|K|$ iterations.

In order to avoid infeasible solution a backtrack procedure has been implemented to ensure the correctness of the algorithm.

Let us indicate with s^j the j-th feasible solution achieved by the matheuristic. In the next iteration, (that is, the next commodity chosen) $\mathcal{M}$ can result unfeasible due to an inconsistent choice of values for fixed variables within V_y and V_z. To recover from an unfeasible state, the backtrack strategy unfixes each variable that was fixed after computing s^j and $\mathcal{M}$ is solved again. If unfeasibility still arises, the same is done with respect to s^i and $i = j - 1$. The process is iterated up to $i = 1$, case in which $\mathcal{M}$ is solved with all unfixed variables in V_y and V_z.

After the $|K|$ iterations, the integrality for all the variables left in V_y^r and V_z^r is restored and all the subsets are coherently updated. In order to have a mathematical formulation simpler to solve, a further fixing step is performed on binary variables in V_z. Let $\bar{z}_{it}$ be the values of variables z_{it} obtained in the last solution for each $i \in Q$ and $t \in T$. Parameter $\theta \in [0, 1]$ is defined to compare its value with each $\bar{z}_{it}$ and variable z_{it} is fixed to one if $\bar{z}_{it} \geq \theta$. $\mathcal{M}$ is then solved and a feasibility check is performed again. If the current state is unfeasible, θ is incremented by a small step α and all the z_{it} for which $\bar{z}_{it} < \theta$ are unfixed; then $\mathcal{M}$ is solved again. This is done until $\mathcal{M}$ is feasible or it results unfeasible for $\theta = 1$, case in which the backtrack strategy is applied up to a final feasible integer solution. For each commodity $k \in K$ that is not purchasable by an external supplier, $c_k = \infty$ and this is assumed to preserve the consistency within (13). This selection prioriry, that is, selecting those commodities in later steps, helps to prevent unfeasibility, as variables f_t^k are enforced to be zero and cannot balance the possible increment of requests r_{it}^{kh}, due to the selection of device i in further steps of the algorithm.

4.2   Time-Based Scheme $\mathcal{H}_T$

The second scheme $\mathcal{H}_T$ implemented is based on considering subsets of timesteps of time horizon and defining the operative phase of the energy system in subsequent stages of decisions. The design choices are implied from the results of the operative schedule.

Recall the notation and the initialization introduced in Sect. 4.1, and let us further introduce the auxilary set T_r initialized as the empty set. A fixed number τ of timesteps is selected from T according to different criteria. Firstly, for each commodity $k \in K$ timesteps $t_{max}, t_{min} \in T$ are added to T_r and are defined as

$$t_{max} = \arg \max_{t \in T \setminus T_r} \{d_t^k\} \quad t_{min} = \arg \max_{t \in T \setminus T_r} \{d_t^k\}$$

respectively; that is, a timestep t is selected if among all the $t \notin T_r$ the maximum (minimum) demand of commodity k is reached at t. Moreover, the mean demand d_{avg}^k is computed for each $k \in K$ and $t_{avg} \in T$, defined as $t_{avg} = \arg \max_{t \in T \setminus T_r} |d_{avg}^k - d_t^k|$, is added to T_r, so that the absolute difference from the average demand is minimized. Finally, a number of additional timesteps are randomly extracted from T and inserted in T_r, until $|T_r|$ is multiple of τ.

Once T_r is populated, for each $i \in Q \setminus S$ and $t \in T_r$ the integrality is restored for variables z_{it} that are removed from V_z^r and moved to V_z. Formulation $\mathcal{M}$ is then solved and all variables z_{it} are fixed to their integer values for each $t \in T_r$. Notice that, due to constraints (5), a variable y_i is implicitly enforced to be 1 whenever a z_{it} is fixed to 1 for at least a $t \in T_r$. The whole process is iterated until a sufficiently large subset of elements of T also belongs to T_r, that is $\frac{|T_r|}{|T|} \geq \Delta$ with $\Delta \in [0,1]$. The same backtrack strategy described in Sect. 4.1 is called to manage unfeasible solutions.

The integrality is then restored for all the variables up to empty subsets V_y^r and V_z^r. Similarly to $\mathcal{H}_K$, the fixing step based on the evaluation of threshold θ is used to reduce $|V_z|$.

The performance of $\mathcal{H}_T$ are significantly affected by the choice of parameter τ. Intuitively, for smaller values of τ a limited number of binary variables is added each time in $\mathcal{M}$ and the computational burden to solve the formulation is (possibly) lower; however, the quality of the solution generally degrades as the integrality constraints are considered only for small subsets of variables. The opposite usually happens for larger values of τ and this guess has been confirmed in the computational experience.

5   Computational Experiments

The code was implemented in AMPL v.20180308 (MS VC++ 10.0, 64-bit) and experiments were performed on a Intel® Core i7-960 3.20 GHz with 6 Gb RAM. All the MILP were solved by IBM® CPLEX® 12.5.0.0.

The performance of both $\mathcal{H}_K$ and $\mathcal{H}_T$ were evaluated and compared on a group of nine instances derived from the real case studies and a group of fifty randomly generated instances, all based on the energy hub scheme depicted in Sect. 2: 4 energy vectors are taken into account, each one uniformly measured in kWh; the set Q is composed by 55 technologies, among which 37 conversion appliances and 18 storage devices. For conversion technologies, the lower bound of produced energy is set equal to 30% of the maximum rated power. After a thorough tuning, parameter ϵ was set to 0.1 for $\mathcal{H}_K$, Δ was chosen as 0.5 for $\mathcal{H}_T$ and $\theta = 0.6$ and $\alpha = 0.1$ were selected for both approaches. Parameter τ was specified differently for the two groups of instances. Finally, it is worth mentioning that feasible integer solutions were obtained in all the instances without relying on the backtrack procedure, so that the fixing strategies emerged sufficiently flexible to avoid unfeasible states.

5.1   Randomly Generated Instances

The group of fifty random instances was built by randomly generating the values of demands within [0, 250] kWh for each time unit $t \in T$, while a real profile was exploited to derive the solar radiation. The time horizon length was set equal to 48 time steps, corresponding to a period of two days in which the demand of users must be met. Parameter τ was fixed at 30 for $\mathcal{H}_T$.

In Table 1 are reported the mean results for five subgroups, each one made by ten instances, achieved by the matheuristic with the two schemes $\mathcal{H}_K$ and $\mathcal{H}_T$, respectively. In particular, for each scheme is given the optimality gap obtained with respect to the optimal solution provided by CPLEX when used to exactly solve the MILP formulation $(\mathcal{M})$. Additionally, the CPU time required by each scheme and by CPLEX is reported.

Both the frameworks reaches very limited optimality gaps, always not larger than 0.60% on the mean. Specifically, on the average $\mathcal{H}_K$ achieves a gap of 0.29%, with a worst value of 1.11% on a single instance; for $\mathcal{H}_T$ instead, the mean gap reached across all the subgroups is of 0.38%, with a worst case of 1.44% on a single instance. The CPU time spent by $\mathcal{H}_K$ and $\mathcal{H}_T$ is largely smaller than the average time required by CPLEX to solve $\mathcal{M}$. The heuristics run in about four minutes on the mean, whereas the solver needs roughly an hour to find an optimal solution. The CPU time spent by $\mathcal{H}_T$ defines a standard deviation of 167.62, whereas for $\mathcal{H}_K$ the standard deviation is equal to 660.62, thus highlighting a much wider spread of the values. Indeed, this can be ascribed to three instances in which $\mathcal{H}_K$ was particularly inefficient and required more than 1400 s.

Summarizing, $\mathcal{H}_K$ obtained slightly better optimality gaps values with respect to $\mathcal{H}_T$, although $\mathcal{H}_T$ appeared faster and especially more stable in terms of the computational effort required to solve the instances.

Table 1. Solutions for random instances.

Instance	$\mathcal{H}_K$		$\mathcal{H}_T$		$\mathcal{M}$
	gap%	CPU	gap%	CPU	CPU
I_1	0.44	561.92	0.28	212.47	3494.70
I_2	0.29	25.03	0.60	141.05	2603.54
I_3	0.28	143.22	0.36	152.84	3656.98
I_4	0.13	83.93	0.30	193.77	4792.81
I_5	0.33	384.81	0.36	177.8	4618.19
Average	0.29	239.78	0.38	175.59	3833.25

5.2 Case Studies

The proposed method is validated by means of two realistic case studies consisting in annual commodity demands and solar radiation availability with hourly resolution: a residential test case represented by real measured data, and a tertiary test case (office buildings) represented by simulated data. The form test case is obtained from a publicly available dataset, which contains measured consumption data for a series of households [26] in the United States. Such test case is built by aggregating the consumption of fifty households within a single yearly demand curve per each considered commodity: electricity, space heating and space cooling. The latter test case is built by computing yearly demands for the same three commodities from office building models developed by the U.S. Department of Energy [27], such models are then simulated by means of the software Energy Plus [28]. The simulations are set considering the climate of a city situated in central Italy. For both locations, the solar radiation is given with the same yearly distribution with hourly resolution as for the demands, with the data coming from the databases of the Energy Plus software that represents an average year obtained from historical data for each of the two sites. The availability of the solar radiation resource is thus limited within each single timestep by the value of the representative parameter. Both the two district test cases are considered to be set in urbanized settings, thus with access to external distribution infrastructures such as an electricity and natural gas distribution networks, from which the respective commodities can be purchased in unlimited quantities at each timestep.

5.3   Realistic Instances

From the two case studies a group of nine realistic instances was derived:

– seven instances with a time horizon of 2160 consecutive hours (3 months), among which four (i_1, i_2, i_3, i_4) built from the data of the tertiary test case and the remain three (i_5, i_6, i_7) from the profiles of the residential test case. The time horizons were selected either to span within a single season or to cross between subsequent season, in order to capture different trends of demand. The detail of the real measured demands is shown in kWh in Fig. 2; the availability of the solar resource for a surface with an inclination of 30° in kW/m^2 is shown in Fig. 3.
– The remaining two instances based on the whole annual demands aggregated into a daily resolution, j_1 related to the tertiary test case and j_2 to the residential one. The time horizon is translated into 365 days, so that the size of the MILP remains contained and the instances gives a reasonable coarse-grained description of the district behaviour for the entire year. To preserve the coherence, all technical parameters of systems are scaled within the annual instances.

To solve the realistic instances, an optimality gap tolerance of 2% was set for CPLEX in each iterations of the two versions of the heuristic. Moreover, $\tau = 120$ was used for $\mathcal{H}_T$.

Table 2 summarizes the results obtained by the matheuristic with the schemes $\mathcal{H}_K$ and $\mathcal{H}_T$. For each instances, the total cost Ω and the CPU time (in seconds) are reported.

Since it was not possible to obtain the optimal solution of the original MILP with CPLEX due to the intractability of the formulation with real instances,

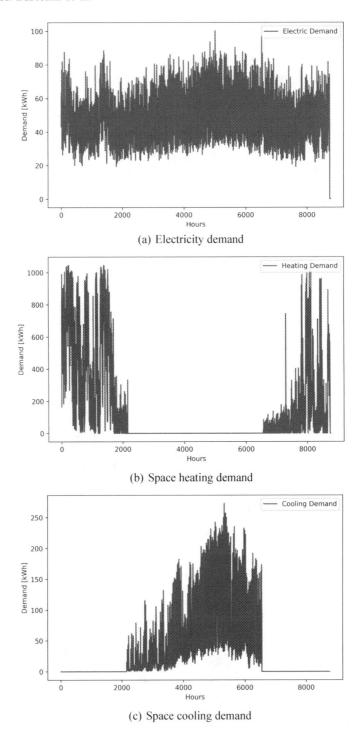

(a) Electricity demand

(b) Space heating demand

(c) Space cooling demand

Fig. 2. Yearly energy demands of the district.

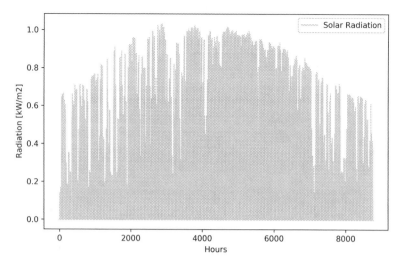

Fig. 3. Yearly solar radiation.

an alternative gap has been computed that is more accurate than a mere linear relaxation of the model. In fact, column $\overline{gap}\%$ reports the percentage gap obtained by comparing the heuristic integer solution with the lower bound resulting by the solution of the MILP in the first step of matheuristic; that is the linear relaxation of $\mathcal{M}$ restricted only by the integrality of the variables selected in the first subset. For $\mathcal{H}_K$, this corresponds to variables y_i related to the technologies that convert (store) the first selected commodity, whereas for $\mathcal{H}_T$ the subset is made by the z_{it} variables for each timesteps t selected at the first iteration.

Table 2. Solutions for realistic instances.

Instance	$\mathcal{H}_K$			$\mathcal{H}_T$		
	Ω	$\overline{gap}\%$	CPU	Ω	$\overline{gap}\%$	CPU
i_1	31957	10.33	1031.32	33593	16.83	1532.68
i_2	26695	13.08	2529.92	27831	15.74	2836.19
i_3	27176	14.09	855.57	28125	16.42	930.83
i_4	30991	18.49	796.42	32436	23.55	602.87
i_5	32249	9.04	1193.94	32874	10.73	925.04
i_6	23300	27.51	9303.47	23535	29.49	10456.34
i_7	20671	27.00	4247.19	19192	16.81	1657.06
j_1	87789	10.55	68.11	87443	6.59	783.28
j_2	68552	12.67	1139.98	70204	10.14	599.04
Average	-	15.86	2351.77	-	16.26	2258.15
Total	349379	-	21165.91	355232	-	20323.33

On the group of the realistic instances, both heuristic approaches obtain values of $\overline{gap}\%$ for which the maximum distance from the optimal solution is not excessive. $\mathcal{H}_\mathcal{K}$ generally obtains better primal solutions, 1.67% lower on the average with respect to $\mathcal{H}_\mathcal{T}$. However, the lower bound provided by $\mathcal{H}_\mathcal{T}$ is on the mean 2.07% better when compared to the one obtained by means of $\mathcal{H}_\mathcal{K}$. In three cases (i_7, j_1, j_2) the primal bound of $\mathcal{H}_\mathcal{T}$ is better than the primal bound of $\mathcal{H}_\mathcal{K}$. Comparing the optimality gaps, on the mean $\mathcal{H}_\mathcal{K}$ achieves a $\overline{gap}\%$ of 15.86% which is slightly smaller than 16.26% obtained by $\mathcal{H}_\mathcal{T}$. The CPU time spent by $\mathcal{H}_\mathcal{K}$ and $\mathcal{H}_\mathcal{T}$ is 2351.77 s and 2258.15 s on the mean, respectively, so that $\mathcal{H}_\mathcal{T}$ results 4.15% faster on the total mean. Given the dimension of the instances, the time required to solve the problem is acceptable. In particular, solutions to the problem of design and management of MES support the decision in a mid-long perspective for which long term planning horizon are considered. Thus, the running time spent by $\mathcal{H}_\mathcal{K}$ and $\mathcal{H}_\mathcal{T}$ are reasonable when compared to exact model-based methods that would ask much longer computational time.

6   Conclusion and Future Steps

In this paper we extended the contents of [14] in which a MILP formulation for the optimal design of energy hubs was proposed, with the aim of minimizing the costs for meeting the energy demands of an community. The proposed model wishes to take into account the synergies between energy vectors, so to exploit potential opportunities in energy savings (and thus costs) that these entail. Given the size of the problem, its solution by means of a MILP is prohibitive and limited only to small instances. Thus a matheuristic approach is proposed, which is based on fixing the values of subsets of binary variables while other integrality constraints are relaxed, iteratively moving up to a feasible integer solution. This is done accordingly to two different strategies: $\mathcal{H}_\mathcal{K}$ and $\mathcal{H}_\mathcal{T}$; the former is based on commodity decomposition and the latter on time-based decomposition.

Results show that both $\mathcal{H}_\mathcal{K}$ and $\mathcal{H}_\mathcal{T}$ can reach very low optimality gaps on small instances by spending a limited amount of time. When moving to realistic instances of interesting size, limited gaps are achieved with respect to computed lower bounds in reasonable time. Both $\mathcal{H}_\mathcal{K}$ and $\mathcal{H}_\mathcal{T}$ can be further seen as methods to provide good primal solutions to feed MILP solvers as CPLEX and solve the instances to optimality, while also giving a first measure of the optimality gap.

Further work will involve the definition of a more sophisticated resolution algorithm, where the mathematical formulation $\mathcal{M}$ is splitted into two models able to interact and correct each other in order to converge to an optimal solution. A possible division could be done separating the investment decisions from the operative schedule of technologies and supplies of energy vectors bought from the external network. Moreover, different heuristic approaches can be implemented in order to obtain a primal bound (i.e., a feasible solution) without using a commercial solver.

References

1. European Commission, Roadmap 2050. Technical report, April 2012
2. BNEF: New energy outlook 2017, 6 (2017)
3. European Commision: JRC science hub: PV status report 2017. Technical report (2017)
4. Fraunhofer ISE: Current and future cost of photovoltaic, long-term scenarios for market development, system prices and LCOE of utility-scale PV systems. Technical report (2015)
5. California ISO: What the duck curve tells us about managing a green grid. Technical report (2012)
6. Jülch, V.: Comparison of electricity storage options using levelized cost of storage (LCOS) method. Appl. Energy **183**, 1594–1606 (2016)
7. Speer, B., et al.: The role of smart grids in integrating renewable energy. Technical report (2015)
8. Enea: Urban microgrids. Technical report, pp. 1–55 (2017)
9. Center for Climate and Energy Solutions: Microgrids: what every city should know. Technical report (2017)
10. Mancarella, P.: MES (multi-energy systems): an overview of concepts and evaluation models. Energy **65**, 1–17 (2014)
11. Lund, H., Østergaard, P.A., Connolly, D., Mathiesen, B.V.: Smart energy and smart energy systems. Energy **17**, 556–565 (2017)
12. Jana, K., Ray, A., Majoumerd, M.M., Assadi, M., De, S.: Polygeneration as a future sustainable energy solution–a comprehensive review. Appl. Energy **202**, 88–111 (2017)
13. Pfenninger, S.: Dealing with multiple decades of hourly wind and PV time series in energy models: a comparison of methods to reduce time resolution and the planning implications of inter-annual variability. Appl. Energy **197**, 1–13 (2017)
14. Bartolini, A., Comodi, G., Marinelli, F., Pizzuti, A., Rosetti, R.: A matheuristic approach for resource scheduling and design of a multi-energy system. In: Proceedings of ICORES 2019, pp. 451–458 (2019). ISBN 978-989-758-352-0
15. Escudero, L.F., Salmeron, J.: On a fix-and-relax framework for a class of project scheduling problems. J. Ann. Oper. Res. **140**, 163–188 (2005)
16. Triadó-Aymerich, J., Ferrer-Martí, L., García-Villoria, A., Pastor, R.: MILP-based heuristics for the design of rural community electrification projects. Comput. Oper. Res. **71**, 90–99 (2016)
17. Singh, B., Sharma, J.: A review on distributed generation planning. Renew. Sustain. Energy Rev. **76**, 529–544 (2017)
18. Mehleri, E.D., Sarimveis, H., Markatosa, N.C., Papageorgiou, L.G.: A mathematical programming approach for optimal design of distributed energy systems at the neighbourhood level. Energy **44**, 96–104 (2012)
19. Omu, A., Choudhary, R., Boies, A.: Distributed energy resource system optimisation using mixed integer linear programming. Energy Policy **61**, 249–266 (2013)
20. Bischi, A., et al.: A detailed MILP optimization model for combined cooling, heat and power system operation planning. Energy **74**, 12–26 (2014)
21. Elsido, C., Bischi, A., Silva, P., Martelli, E.: Two-stage MINLP algorithm for the optimal synthesis and design of networks of CHP units. Energy **121**, 403–426 (2017)
22. Li, B., Roche, R., Paire, D., Miraoui, A.: Sizing of a stand-alone microgrid considering electric power, cooling/heating, hydrogen loads and hydrogen storage degradation. Appl. Energy **205**, 1244–1259 (2017)

23. Sachs, J., Sawodny, O.: Multi-objective three stage design optimization for island microgrids. Appl. Energy **165**, 789–800 (2016)
24. Mohammadi, M., Noorollahi, Y., Mohammadi-ivatloo, B., Yousefi, H.: Energy hub from a model to a concept–a review. Renew. Sustain. Energy Rev. **80**, 1512–1527 (2017)
25. Blaauwbroek, N., Nguyen, P.H., Konsman, M.J., Shi, H., Kamphuis, R.I.G., Kling, W.L.: Decentralized resource allocation and load scheduling for multicommodity smart energy systems. IEEE Trans. Sustain. Energy **6**, 1506–1514 (2015)
26. Pecan Street Smart Grid Demonstration Project: Pecan Street Final Technology Performance Report 2015. https://www.smartgrid.gov/files/Pecan-Street-SGDP-FTR_Feb_2015.pdf
27. U.S. department of Energy: Commercial prototype building models
28. U.S. Department of Energy: Energy Plus v8.9.0

Author Index

Printed in the United States
By Bookmasters